The Developing Microbiome

Lessons From Early Life

The Developing Microbiome

Lessons From Early Life

Edited by

Erika C. Claud

Departments of Pediatrics and Medicine, Sections of Neonatology and Gastroenterology, The University of Chicago, Chicago, IL, United States

ELSEVIER

ACADEMIC PRESS

An imprint of Elsevier

Academic Press is an imprint of Elsevier
125 London Wall, London EC2Y 5AS, United Kingdom
525 B Street, Suite 1650, San Diego, CA 92101, United States
50 Hampshire Street, 5th Floor, Cambridge, MA 02139, United States
The Boulevard, Langford Lane, Kidlington, Oxford OX5 1GB, United Kingdom

Notices
Knowledge and best practice in this field are constantly changing. As new research and experience broaden our understanding, changes in research methods, professional practices, or medical treatment may become necessary.

Practitioners and researchers must always rely on their own experience and knowledge in evaluating and using any information, methods, compounds, or experiments described herein. In using such information or methods they should be mindful of their own safety and the safety of others, including parties for whom they have a professional responsibility.

To the fullest extent of the law, neither the Publisher nor the authors, contributors, or editors, assume any liability for any injury and/or damage to persons or property as a matter of products liability, negligence or otherwise, or from any use or operation of any methods, products, instructions, or ideas contained in the material herein.

British Library Cataloguing-in-Publication Data
A catalogue record for this book is available from the British Library

Library of Congress Cataloging-in-Publication Data
A catalog record for this book is available from the Library of Congress

ISBN: 978-0-12-820602-7

For Information on all Academic Press publications
visit our website at https://www.elsevier.com/books-and-journals

Publisher: Stacy Masucci
Editorial Project Manager: Pat Gonzalez
Production Project Manager: Swapna Srinivasan
Cover Designer: Matthew Limbert

Typeset by MPS Limited, Chennai, India

Contents

List of contributors

Emma Allen-Vercoe Department of Molecular and Cellular Biology, University of Guelph, Guelph, ON, Canada

Supinda Bunyavanich Division of Allergy and Immunology, Department of Pediatrics, Icahn School of Medicine at Mount Sinai, New York, NY, United States; Icahn Institute for Data Science and Genomic Technology, Department of Genetics & Genomic Sciences, Icahn School of Medicine at Mount Sinai, New York, NY, United States

Mary M. Buschmann Department of Pediatrics, University of California San Diego, La Jolla, CA, United States

Stanley Cho Section of Pediatric Gastroenterology, Texas Children's Hospital, Baylor College of Medicine, Houston, TX, United States

Erika C. Claud Departments of Pediatrics and Medicine, Sections of Neonatology and Gastroenterology, The University of Chicago, Chicago, IL, United States

Modupe O. Coker Department of Oral Biology, Rutgers School of Dental Medicine, Newark, NJ, United States; Department of Epidemiology, Geisel School of Medicine at Dartmouth, Hanover, NH, United States

Promi Das Department of Pediatrics, University of California San Diego, La Jolla, CA, United States

Jack A. Gilbert Department of Pediatrics, University of California San Diego, La Jolla, CA, United States

Amy B. Hair Section of Neonatology, Texas Children's Hospital, Baylor College of Medicine, Houston, TX, United States

Hsi-en Ho Division of Allergy and Immunology, Department of Pediatrics, Icahn School of Medicine at Mount Sinai, New York, NY, United States

Stephanie Jeong Division of Allergy and Immunology, Department of Pediatrics, Icahn School of Medicine at Mount Sinai, New York, NY, United States

Richard Kellermayer Section of Pediatric Gastroenterology, Texas Children's Hospital, Baylor College of Medicine, Houston, TX, United States; USDA/ARS Children's Nutrition Research Center, Houston, TX, United States

Mahesh Krishna Rice University, Houston, TX, United States

Jing Lu Department of Pediatrics, The University of Chicago, Chicago, IL, United States

Juliette C. Madan Department of Epidemiology, Geisel School of Medicine at Dartmouth, Hanover, NH, United States; Department of Pediatrics, Dartmouth Hitchcock Medical Center, Lebanon, NH, United States

Camilia R. Martin Department of Neonatology, Beth Israel Deaconess Medical Center, Boston, MA, United States; Harvard Medical School, Boston, MA, United States

Josef Neu Division of Neonatology, Department of Pediatrics, University of Florida, Gainesville, FL, United States

Kaitlyn Oliphant Department of Pediatrics, The University of Chicago, Chicago, IL, United States

Laura Patton Division of Neonatology, Department of Pediatrics, University of Florida, Gainesville, FL, United States

Anita Rao Section of Gastroenterology, Hepatology, and Nutrition, Department of Pediatrics, The University of Chicago, Chicago, IL, United States

Simone Renwick Department of Molecular and Cellular Biology, University of Guelph, Guelph, ON, Canada

Mariana C. Salas Garcia Department of Pediatrics, University of California San Diego, La Jolla, CA, United States

Gregory C. Valentine Division of Neonatology, Seattle Children's Hospital, University of Washington, Seattle, WA, United States

Annie Z. Wang Department of Pediatrics, Section of Neonatology, The University of Chicago, Chicago, IL, United States

Alyson L. Yee Interdisciplinary Scientist Training Program, University of Chicago, Chicago, IL, United States

Preface

The human microbiome is an ecosystem comprised of all the microbes in and on the human body including bacteria, fungi, archaea, and viruses. It functions as a community in which diversity, balance, and functional redundancy are associated with health and resistance to disease. There are equal numbers of microbial and human cells, with the microbes contributing significant mass, genetic material, and metabolic capacity. Therefore to understand human health, one needs to understand microbial community health. These are complex relationships that require understanding beyond the presence or absence of specific organisms or the categories of pathogens versus commensals. Advances in sequencing technology, bioinformatics pipelines, metabonomics, proteomics, and cultivation techniques have given investigators the tools to establish a more robust understanding of the function of the microbiome in health and disease.

The microbiome is influenced by environment, and it is key for human development. Importantly the microbiome develops over time. It may begin in utero; it expands rapidly after birth through contact with the physical environment, caretakers and objects, diet, and therapeutic interventions such as antibiotics; and it undergoes succession to converge on adult patterns by age 3. After that time point interventions such as dietary change, antibiotics, or probiotics may cause fluctuations, but the majority of microbiome patterns will revert to the native community.

There is mounting evidence that short- and long-term health and clinical outcomes are greatly influenced by early life events. The Barker hypothesis notes that low birth weight associated with the intrauterine environment is inversely correlated with increased early death secondary to coronary heart disease [1]. The thrifty phenotype hypothesis suggests that diabetes and metabolic syndrome result from an energy balance regulatory system that has been programmed by in utero maternal nutrient restriction and encounters caloric excess postnatally [2]. Furthermore, environmental, dietary, and iatrogenic factors can disrupt developmental processes dependent on proper succession of the early life gut microbiome such as metabolism, brain and neurological growth, and immune function. The microbiome has been particularly important for improving understanding of conditions that are multifactorial with environmental components, including conditions as disparate as prematurity [3,4], cancer [5], and Alzheimer's disease [6]. Alterations in the early microbiome have also been linked to inflammatory bowel disease (IBD) [7], type 1 diabetes [8], obesity [9], and asthma [10].

Infancy is increasingly being recognized as a key time point of microbiome establishment that impacts neonatal health as well as long-term adult outcomes. This places study of the infant and childhood microbiome in a very important position. While much of medicine and research is performed in silos—obstetrics separate from pediatrics separate from medicine separate from geriatrics—life is a continuum, and early events have a significant impact on later life stages. Understanding of normal microbiome succession and the early childhood events that influence composition are critical not only for infant and childhood health, but also for understanding health and disease across the life course.

This book will explore the microbiome of early life to give an in-depth review of microbiome facets of importance to infant health. It begins with the controversy surrounding the possibility of an in utero microbiome. Next is an examination of environmental influences on microbiome succession of the infant and immune development. Subsequent chapters delve into the role of the microbiome in several specific conditions including prematurity, IBD, neurodevelopment, food allergy, and obesity/malnutrition. It concludes with an exploration of means to influence the microbiome including vitamin D and microbiome-based therapeutics. Each chapter intentionally includes a section on relevance beyond the pediatric patient population in hopes of prompting discussion, research and discovery of the interactions of the microbiome, and health across the life course.

References

[1] Barker DJ. The origins of the developmental origins theory. J Intern Med 2007;261 (5):412–17.

[2] Hales CN, Barker DJ. The thrifty phenotype hypothesis. Br Med Bull 2001;60:5–20.

[3] Wang Y, Hoenig JD, Malin KJ, Qamar S, Petrof EO, Sun J, et al. 16S rRNA gene-based analysis of fecal microbiota from preterm infants with and without necrotizing enterocolitis. ISME J 2009.

[4] Fettweis JM, Serrano MG, Brooks JP, Edwards DJ, Girerd PH, Parikh HI, et al. The vaginal microbiome and preterm birth. Nat Med 2019;25(6):1012–21.

[5] Gopalakrishnan V, Helmink BA, Spencer CN, Reuben A, Wargo JA. The influence of the gut microbiome on cancer, immunity, and cancer immunotherapy. Cancer Cell 2018;33 (4):570–80.

[6] Kim MS, Kim Y, Choi H, Kim W, Park S, Lee D, et al. Transfer of a healthy microbiota reduces amyloid and tau pathology in an Alzheimer's disease animal model. Gut 2020;69 (2):283–94.

[7] Pittayanon R, Lau JT, Leontiadis GI, Tse F, Yuan Y, Surette M, et al. Differences in gut microbiota in patients with vs without inflammatory bowel diseases: a systematic review. Gastroenterology 2020;158(4):930–46.

[8] Livanos AE, Greiner TU, Vangay P, Pathmasiri W, Stewart D, McRitchie S, et al. Antibiotic-mediated gut microbiome perturbation accelerates development of type 1 diabetes in mice. Nat Microbiol 2016;1(11):16140.

[9] Mischke M, Arora T, Tims S, Engels E, Sommer N, van Limpt K, et al. Specific synbiotics in early life protect against diet-induced obesity in adult mice. Diabetes Obes Metab 2018;.

[10] Stokholm J, Blaser MJ, Thorsen J, Rasmussen MA, Waage J, Vinding RK, et al. Maturation of the gut microbiome and risk of asthma in childhood. Nat Commun 2018;9 (1):141.

Chapter 1

Microbes and the womb: does it matter?

Laura Patton and Josef Neu
Division of Neonatology, Department of Pediatrics, University of Florida, Gainesville, FL, United States

Introduction

As more information is discovered about the impact of both pathogenic and beneficial microbes to human health, the question emerges of when the interaction between microbes and humans begins. Furthermore, does the timing of this interaction play a role in subsequent health? One line of thinking is that the newborn's first exposure to microbes is after birth and that the womb from which the infant emerges is sterile. Another theory is that exposure to microbes occurs in the womb before a baby is born. These opposing theories have become a topic of intense debate between those who believe in a sterile womb and those who believe that we at birth already have a distinct microbiome that is acquired in utero [1]. Of note is that these opposing theories portend much more than a simple academic argument. They represent paradigm shifting science with major clinical implications that range from the role of in utero microbes in preterm labor, the development of immune responses including autoimmunity, as well as epigenetically acquired traits that can be passed from one generation to the next via vertical microbial transfer between generations.

In this chapter, we will: (1) provide a historical perspective of the "sterile womb" hypothesis and how technologies for detection and characterization of microbes in various niches such as the vagina, amniotic cavity, placenta, and fetal gastrointestinal tract have altered our perspective; (2) review the literature for and against the "sterile womb" hypothesis and elaborate on studies of microbes in these niches and on how these microbes might colonize or expose the in utero environment to perturbations with long-lasting effects; (3) discuss the relative merits for the opposing positions of sterility of the womb based on existing studies; and (4) discuss the impact of this research to human health.

The Developing Microbiome. DOI: https://doi.org/10.1016/B978-0-12-820602-7.00001-5

Historical perspective and technological developments

Henri Tissier, at the Paris, France Pasteur Institute, is said to have first proposed that the womb was sterile [2]. Tissier found that breastfed infants' stools contained a high quantity of a bacterium that was characterized by a Y-shaped ("bifidus") morphology. He observed that infants who ingested milk containing this microbe had a lower incidence of diarrhea. He inferred that the differences in the enteral bacteria in these babies were due to acquisition after birth, with the in utero environment being otherwise sterile. Exactly how this led to the dogma of the womb being sterile (the "sterile womb hypothesis") and its persistence for over 100 years remains an enigma. However, there are several lines of emerging evidence that challenge this unsubstantiated theory.

Challenges to the "sterile womb" concept have emerged partially due to a greater emphasis on studies of the microbial environment of the womb as well as technologic advances for bacterial isolation and identification. Historically, the most commonly used method of identifying bacteria utilized culturing techniques, which have some major shortcomings. It is a common misunderstanding that microorganisms cultured from an environment represent the most dominant and/or the most significant taxa in that environment. This is incorrect: microorganisms isolated using standard culture-based methods are in reality isolated because of their ability to grow efficiently into colonies under a typical set of conditions—usually under aerobic conditions, at moderate temperatures and with high-nutrient growth media. Thus these easily isolated organisms are the "weeds" of the microbial world [3]. They are estimated to constitute less than 1% of all microbial species in a large variety of microbial samples. Thus examination under a microscope can show various bacteria that will not be identified upon routine culturing. This has been termed the "great plate count anomaly" and was first described in the 1980s [4].

Within this concept, it was recognized that our ability to replicate optimal conditions for growth of bacteria is so limited that the vast majority of bacteria are elusive to when cultured by routine methods. Thus defining an environment or a specimen as truly sterile is very challenging. One can readily but mistakenly presume sterility because of this inability to culture microbes from a certain environment.

Within the past few decades, newer methods of microbial identification have become widely used. Polymerase chain reaction (PCR) testing, which utilizes identification of segments of microbial DNA after amplification, can be very sensitive and has become a new standard for rapid identification of microbes, particularly for those microbes that are difficult to culture. This becomes especially useful when trying to examine specimens collected from sites that are likely to have low quantities of bacteria and/or have been considered sterile, such as cerebrospinal fluid, amniotic fluid, and meconium.

PCR testing will amplify a section of the bacterial or viral genetic code and will allow us to compare it to already recognized genetic sequences to allow for identification.

Another technological advance employs the 16S rRNA gene [5]. This is a gene found in all bacteria, with both highly conserved and variable regions. The variable regions' sequences are utilized to identify the relative abundances of various microbial taxa in a certain sample, evaluate their diversity, and make comparisons to other samples. In another closely related technique, commonly termed whole metagenome sequencing, which analyzes the whole genome of microbes in a certain niche, much more information about the genetic potential of the microbial community can be determined. The latter technique still remains relatively expensive when compared to the 16S rRNA technique, but it provides considerably more information. An extensive review of these techniques is beyond the scope of this review, but it is important to note that these "next-generation" technologies derived from the Human Microbiome Project [6], along with emerging bioinformatics techniques [7], are being actively utilized for microbial characterization of the vaginal, amniotic fluid, placental, and meconium microbiomes, all critical in our discussion of the microbial ecology of the womb.

Origin of opposing views on the "sterile womb hypothesis"

Microbes in the female reproductive tract

In order to obtain perspective on the microbial environment of the female reproductive tract, we will first discuss developments using various analytic techniques available during different epochs that have been utilized to evaluate microbes in different parts of the reproductive system. We will start with the vagina that clearly exhibits microbial colonization, which occurs when microorganisms live on or in a host organism, but do not invade tissue or cause damage.

In 1982 the German gynecologist Albert Doderlein discovered the presence of vaginal microbes that were a source of secretions that had the ability to inhibit the growth of pathogens associated with puerperal fever [8]. These bacteria were subsequently found to belong to the genus *Lactobacillus*, and the protective secretions were lactic acid [9].

Despite the proximity of the vagina to the uterus and the known presence of vaginal microbes, questions pertaining to interaction between vaginal microbes and the in utero environment were not scrutinized until the late 1990s [10]. Studies linking vaginal microbes to preterm delivery were expanded with the hypothesis that vaginal microbes ascend through fallopian tubes, translocate through the choriodecidual membranes, and cause amniotic fluid inflammation, which may subsequently trigger preterm labor [10,11].

The concept that microbes might associate with the uterus in a nonpathologic manner required further technologic development.

In the more recent era, a myriad of studies utilizing both culture- and nonculture-based techniques show not only colonization of the vaginal tract by nonpathogenic microbes, but also a dynamic interaction of vaginal microbes with the human host [12] that in healthy nonpregnant women is influenced by sexual activity, menstrual cycle, local microbiota, and ethnicity [13].

The proximal vaginal (introitus) microbiome has little diversity and is mostly dominated by four species of *Lactobacilli*: (1) *L. crispatus*, (2) *L. iners*, (3) *L. jensenii*, and (4) *L. gasseri* [13]. Recently, Chen et al. [14] evaluated proximal to distal regions of the reproductive tracts of 110 women and examined the nature of colonization using both sequencing and cultivation techniques. These studies found distinct microbial communities in the cervical canal, uterus, fallopian tubes, and peritoneal fluid, differing significantly in diversity and microbial taxonomy from that of the proximal vagina, supporting nonsterility of at least those components of the more proximal reproductive tract.

Microbes in amniotic fluid

Going beyond the vaginal tract and fallopian tubes, research done in humans several decades ago demonstrated that, even without ruptured membranes, bacteria are often found in amniotic fluid and that these were often associated with morbidity to both the mother and offspring. Because many cesarean deliveries are performed prior to rupture of the amniotic sac, this method of delivery provided a new avenue to obtain amniotic fluid that had not been exposed to the external environment and thus to test the sterility of the in utero environment. If performed carefully, surgeons could collect a "sterile" sample of amniotic fluid at the time of delivery. This fluid could then be cultured to determine presence or absence of microbes.

Using this method, in 1927, Harris and Brown collected amniotic fluid from healthy term cesarean section deliveries. They found that for all women in labor less than 6 hours, bacterial culture of amniotic fluid was negative. All women in labor longer than 6 hours had positive cultures [15]. Those that produced positive culture results preferentially grew *Staphylococcus albus*, *Streptococcus*, and yeast, making samples' results suspicious for contamination from skin flora. Thus these results propagated the dogma that the amniotic cavity is sterile prior to labor onset.

With further development of techniques for sampling of the amniotic cavity such as amniocentesis, culture-based analysis of amniotic fluid from a small number of cases from elective pregnancy termination recovered Mycoplasma and Staph species, which was suggestive of contamination from skin or vaginal examination [16]. Another study used culture-based methods wherein bacteria were recovered from amniotic fluid in 7 out of 10 cases of spontaneous preterm labor with intact membranes [17]. This study suggested that clinically silent amnionitis was associated with and possibly caused preterm labor. It also suggested a pathogenic role of anaerobic bacteria including *Bacteroides* species, and it was one of the first of several studies relating microbial invasion of the amniotic cavity (MIAC) with preterm delivery.

Additional studies showed that inflammatory responses associated with these intrauterine microbial communities were linked to various pathologies including preterm delivery [18], as well as development of brain, lung, and eye disease in these infants [19,20]. Thus much of the thinking about microbes and the uterine cavity suggested that if microbes were present, they played a pathogenic role, rather than the commensal role.

Concurrent with the initiation of the Human Microbiome Project in 2007, DiGiulio and colleagues evaluated samples of amniotic fluid using both culture- and PCR-based techniques and observed the presence of numerous bacterial taxa in amniotic fluid for the first time. They also found that microbial DNA load was inversely related to preterm birth and directly associated with levels of IL-6 and white blood cells in the amniotic fluid [21,22]. They proposed four potential routes of microbial invasion into the amniotic cavity, which included (1) migration from the vagina through the cervix, (2) transplacental invasion after hematogenous dissemination, (3) seeding from the fallopian tubes, (4) and/or inoculation at the time of amniocentesis or other invasive procedure. The presence of microbes that could be cultured or detected by PCR in the amniotic fluid of mothers who did not exhibit pathology in these studies provided additional evidence that the womb may not necessarily be sterile and that some of the microbes *may not* represent pathogens.

Microbes in placenta

The placenta is thought of as a protective barrier between mother and baby, filtering pathogens to a remarkable degree. The placenta is formed from both fetal and maternal components that merge creating a unique filter that can provide essential nutrients and is thought to stave off many microbial pathogens during fetal development. Several lines of evidence support the concept that the placenta is only invaded by pathogenic microbes, and commensal microbes do not play a role in terms of colonization.

In one study, placentas from both healthy pregnancies and those with suspected infections were studied. While 11 out of 33 placentas in the suspected infection pregnancies had positive bacterial cultures, only 1 out of 46 in the healthy controls had a positive bacterial culture [23]. The marked difference between these groups suggested that the presence of bacteria within the placenta does indeed spark concern about infection rather than simple colonization.

On the other hand, it is also well illustrated that multiple viruses can cross the placental barrier and affect the fetus; for example, viruses that comprise the "TORCH" (toxoplasosis, other, rubella, cytomegalovirus, and herpes) infections are some of the most prominent. Certain bacteria such as *Listeria* are also able to cross this barrier, thus the placental barrier should not be considered as impenetrable [24].

In spite of it being perceived as a vital component in the defense of a fetus, research into the placenta's role in the neonatal or fetal microbiome is relatively new. Again, as already stated, much information had previously been documented about bacterial presence in the placenta during presumed infection but showing benign presence of bacteria was sparse. Whether commensal bacteria are associated with the placenta was more recently addressed.

In 1982 Kovalovski et al. found aerobic bacteria in 16% of placentas, which he studied using culturing techniques [25]. Within this study, there was no notable evidence of placental inflammation or infection. Similarly, Hiller et al. found bacterial presence in 21% of placentas on a study conducted in 1988 [26].

In an evaluation of 195 total patients focusing on placental basal plates that included histologic staining for bacteria, it was found that of the total 195 patients evaluated, intracellular bacteria were detected in a large number of samples without clinical or pathologic chorioamnionitis [27].

Other studies demonstrated the presence of commensal bacteria in healthy placentas using newer technologies including next-generation sequencing techniques. In 2014 a study by Aagaard et al. [28] involved a cohort of placental specimens collected under sterile conditions from 320 subjects. Both 16S ribosomal DNA-based and whole-genome shotgun (WGS) metagenomic studies were employed. Microbial taxa identified in the placenta were compared to other body site niches as well as nonpregnant controls. This study suggested a unique placental microbiome composed of nonpathogenic commensal microbiota from various phyla. Because of its major implications, this study stimulated considerable controversy, including negative remarks about these studies in scientific journals (e.g., "reagent contamination can lead to spurious microbiome descriptions," which "go against basic understanding of microbial ecology, some of which can be likened to reporting blue whales in the Himalayas or African elephants in Antarctica" [29,30]), on social media as well as popular press (e.g., The Atlantic, July 2019). The argument against a placental microbiome was largely based on other studies suggesting that the bacteria identified in the Aagaard study as well as others were due to the lack of adequate controls and reagent contamination—a "kitome" [31,32]. Further concerns were elaborated in a review by Perez-Munoz, wherein other arguments against a placental microbiome were raised, including the existence of germ-free animals, which allegedly refute the placental microbiome hypothesis [33].

This continues to be a matter of contention. In the studies that suggested microbial presence in the placenta, some of the relative abundances of bacteria differed and the placental microbiome indeed differed from DNA extraction and PCR blanks. Even after controlling for delivery method, the resemblance between the vaginal and placental microbiome was still present, supporting that the finding of the microbes in the placenta was not due to contamination from vaginal microbes [34].

The variability in findings from different studies can at least partially be explained by the different sections of placental tissue collected in each experiment. Certainly different layers of placenta could harbor their own unique microbial communities. In fact, the studies that found similarities between the placental and oral microbiome sampled the chorionic villi. Vaginal microbiota were present in samples from the entire placenta, and Enterobacteriaceae were found in decidual samples. Studies also suggest that the fetal membrane and basal plate of the human placenta differ in terms of bacterial abundance and diversity [14].

Although much of the controversy focuses on nonculture-based techniques, viable microbes have been cultured from human placenta, amniotic fluid, and umbilical cord blood samples. One study from China [35] that employed both culture- and nonculture-based techniques on amniotic fluid and placental samples demonstrated that all of the 50 amniotic fluid samples studied were culture-negative, whereas 379 out of 1832 placenta samples were culture-positive. These studies suggested that womb contained microbiome with low diversity, with cultivatable microorganisms in placenta and suggested that exposure to microbes does occur during healthy pregnancy.

These discrepancies have led to a "scientific knife fight" [1] primarily based on social media blogs. Before we discuss this argument further, we will present additional information about microbes in meconium, the baby's first stool

Meconium

Meconium is a dark, viscous material that constitutes the first stool of the newborn. Meconium gives a unique look into the fetal environment. It is produced in utero and develops within the fetus's intestinal tract as it is forming. As the fetus develops, it swallows amniotic fluid which then contributes to the development of meconium during later aspects of fetal development. Being that meconium stays within the intestinal tract throughout this time, it can serve as a marker for environmental exposures such as drugs [36]. Sometimes meconium is expelled when the fetus is under stress around delivery but usually the first meconium is passed within the first day after delivery in full-term neonates. Babies born more prematurely can take longer to pass meconium and this can make evaluation of their meconium more difficult as the delay in passage leaves time for enteral feeding to play a role in microbial communities [37]. One can readily surmise that if meconium contains microbes, then they are either derived from in utero sources or contamination during anal passage.

Theodor Escherich, in Germany, conducted some of the first studies on meconium in the 1880s. His work showed that meconium had no viable bacteria within it, again giving credence to the idea of a sterile womb [38]. Further studies showed that not all samples of meconium collected under sterile conditions in healthy pregnancies were "sterile" [39]. However, in this study, longer times between birth and passage of meconium were directly correlated. This suggested that increasing quantities of bacteria in meconium were caused by exposure of the meconium to the postnatal environment.

The advent of cesarean section deliveries becoming common in the early 20th century provided additional opportunities to study this concept of in utero sterility. Biasucci in 2008 found that meconium tested for bacteria within the first 3 days after birth [40] had a marked difference in the type of bacteria represented when compared to later specimens. Specifically those delivered via cesarean delivery had a marked absence of *Bifidobacteria* species. *Bifidobacteria* is a predominant bacterial species within the meconium of vaginally delivered infants. Its absence in cesarean delivered babies, particularly when no other intrauterine differences are noted between vaginally and cesarean delivered babies, suggested that this variability can only be acquired by method of delivery and not from other causes.

Dominguez-Bello et al. reported an essentially undifferentiated bacterial community across all newborn surfaces: skin, oral, nasopharyngeal, and gastrointestinal (meconium). This study found that the biggest contributor to the type of bacteria present was the mode of delivery. Vaginal deliveries had a predominance of vaginal bacterial flora, whereas cesarean deliveries were colonized by communities resembling the maternal skin flora [41]. Although this does not specifically speak to a sterile intrauterine environment, it suggests predominantly extrauterine sources of bacteria as the major components of the neonatal microbiome when delivered by C-section.

With the advent of nonculture-based bacterial testing such as PCR and 16S rRNA gene testing, more studies are finding earlier presence of bacteria within fetal and neonatal environments. Jimenez et al. reported 100% of mouse meconium samples being positive for bacteria [42]. Multiple studies since have demonstrated a bacterial presence in human preterm infant meconium using PCR or 16S rRNA gene methods in addition to culture-based methods [43,44]. Hansen et al. found viable bacteria in 66% of meconium samples taken from term vaginally delivered neonates [45]. Hansen found a very low quantity of bacteria, but confirmation by more standard culture-based methods was difficult or unable to be performed on most samples. More recent studies have evaluated the intestinal microbes of mice and have found the presence of cultivatable live microbes, refuting the argument for "sterility" and suggesting a "dynamic, viable mammalian fetal microbiota during in utero development" [46].

Potential sources of intrauterine bacterial transmission to the fetus

If there is in fact colonization in the womb, we need to question the source of in utero colonization. We have evidence for three most likely potential sites of microbial origin for the maternal–fetal microbiome (Fig. 1.1). These include (1) the vagina, (2) the maternal oral cavity, and (3) the maternal intestinal tract [47].

A widely accepted concept is that microbes residing in the vagina (a first potential site) ascend to the choriodecidual membranes, where they translocate into amniotic fluid [11,18]. In these studies, it is suggested that an inflammatory response originating from the decidua leads to preterm labor.

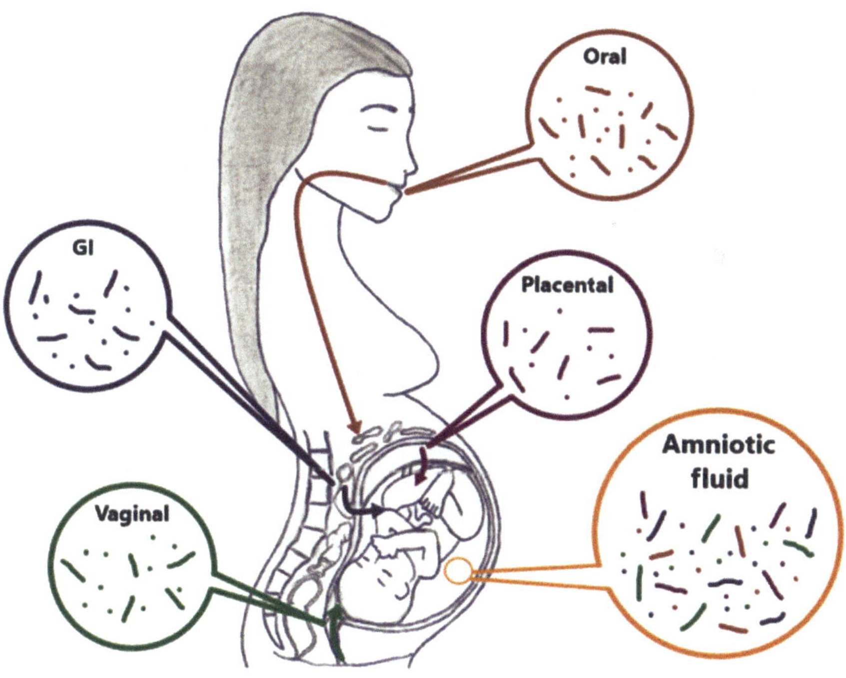

FIGURE 1.1 Methods of maternal–fetal microbial transference, passed into the gastrointestinal tract. The proposed mechanisms of transference to the fetus while within the womb. Oral, gastrointestinal (*GI*), placental, and vaginal microbes are theorized and found across multiple studies as sources of bacteria isolated from amniotic fluid. The fetus swallows this amniotic fluid during development of both the lungs and the intestinal tract. During this process, these bacteria are swallowed and then incorporated into meconium.

Of note, a similarity between microbial taxa found in the placenta and those found in the mouth (derived from a database of nonpregnant women) was discovered. This suggests the mouth as a second site of origin of the maternal–fetal microbiome.

The fact that Escherichia *coli* was one of the largest taxa represented in the placental samples [28] is of interest. These are common residents of the intestine and thus *suggest a third potential site—the maternal intestine.* There is additional data that supports a maternal intestinal origin.

Microbes in the pregnant woman's intestinal tract differ depending on stage of pregnancy. Fecal bacteria from 91 pregnant women during different stages of pregnancy were analyzed in one study in order to determine more specifically the changes that occur during pregnancy [48]. Gut microbiota changed dramatically from first to third trimesters, with an overall increase in Proteobacteria and Actinobacteria, and reduced richness (lower species count). It has been hypothesized that intestinal dysbiosis in the pregnant woman is an emerging cause of pregnancy complications including fetal rejection (causing preterm delivery) [49]. Further support for the possibility of the maternal intestine as the source of fetal–maternal microbes is derived from several studies: In one study, a group of pregnant mice were orally inoculated with a genetically labeled *Enterococcus faecium* strain previously isolated from breast milk of a healthy woman. The labeled strain could be isolated and detected by PCR from the amniotic fluid of the inoculated animals. In contrast, it could not be detected in the samples obtained from a noninoculated control group [50]. In a follow up study, they detected this labeled microbe in meconium of the offspring after feeding to pregnant mice [42]. In another study, it was found that a probiotic administered to pregnant women could alter both placental and fetal intestinal inflammatory markers [51].

The concept of a dormant blood microbiome that distributes microbes, their antigenic components, or their metabolic products between organs such as the maternal intestine and the fetal tissue has also been raised [52]. This is highly significant in that if the maternal intestine is the origin of the microbes (or their components or products) that are found in the fetus, amniotic fluid and placenta, and if these indeed are related to metabolic and inflammatory responses that originate in the maternal and/or fetal intestine that may induce preterm labor, this offers an exciting area for intervention. These include maternal dietary manipulation and microbial therapeutic alterations by providing certain microbes into the mother's diet [53]. Furthermore, employing very specific antimicrobial therapies (e.g., bacteriophages) prevent those microbes associated with preterm labor from thriving in the maternal intestine and consequently the placenta, amniotic fluid, and the fetus.

The role of stress as affecting the intestinal microbiome, hypothalamic function, and intestinal physiology is well known [54–56]. It has also been shown that administering a probiotic formula can reduce anxiety-like behaviors in rats [57]. Furthermore, stress during pregnancy has been related to the microbiome and found to have profound effects on the fetus as well as predisposing to preterm delivery [58,59]. The role of the microbiome as an intermediate in the mechanism of stress induced preterm labor provides another area for intervention. This requires additional studies wherein samples from different sites in the same woman are analyzed in order to determine the origin of the microbes that inhabit the fetal–maternal unit.

It has been dogma that the baby's first stool (meconium) is sterile. This dogma is refuted by studies from our group and others demonstrating that meconium from both preterm and term neonates frequently contain cultured and as yet uncultured microbes [42–44,60,61]. The origin of these microbes is unknown, but tenable support for the hypothesis that they are derived from swallowed amniotic fluid has been bolstered by our studies that show a high degree of similarity between meconium microbes and those found in amniotic fluid [44]. As with the studies on amniotic fluid and placenta, the taxonomy of meconium microbes differ depending on gestational age at the time of delivery, further enhancing a potential role of certain microbes and preterm delivery. We have thus proposed that the fetal intestine, a highly immunoreactive organ [62,63], may represent an alternative to the placenta as a site of origin of this inflammatory response [44].

Arguments for and against a sterile womb: the verdict

"Be curious, not judgmental." Walt Whitman

Numerous arguments for and against the "sterile womb" hypothesis have been raised, some of which appear to show harshness beyond pure scientific inquiry, referred to by one author as a scientific "knife fight" [1]. Here, we will not attempt to adjudicate this issue for one side or the other but rather present several arguments and counterarguments that have been raised and then suggest a roadmap on how to proceed.

There are several arguments raised for and against the sterile womb hypothesis (summarized in Table 1.1):

1. Studies in numerous invertebrate and vertebrate species support maternal transmission of microbes from mother to offspring [64]. It is proposed by the authors reviewing these transmission patterns in various species that the "presence of maternal transmission at the base of the Animalia kingdom and the surprising plasticity by which microbes gain access to germ cells or embryos in these systems signifies that maternal symbiont transmission is an ancient and evolutionarily advantageous mechanism inherent in animals, including humans" [64].
2. From teleological and evolutionary perspectives there may be advantages to the fetus being exposed to microbes. Here it is important to

TABLE 1.1 "The sterile womb" controversy.

For	Against
Very low biomass identified, difficult to claim colonization.	Invertebrate and vertebrate species showing maternal transmission of microbes to offspring.
Ability to generate germ-free mice refutes the concept of microbes within the placenta.	Lack of significant immune response after delivery.
Very few studies show live microbes in placental tissue.	Commensal microbes preventing pathogenic microbial colonization (colonization resistance).
The problem of low biomass samples showing controls that are similar—a "kitome."	Upper female reproductive tract is not sterile, likely with exposure to the uterus.
The problem of defining colonization versus simple temporary exposure.	Animal studies showing microbes within brain, placenta, fetal intestine.
	At delivery, the fetus can produce IgM via plasma cells, promoting a preprogrammed immune system.

note that advantages include immune education with early tolerance to commensal microbes. If this did not exist, then the sudden exposure to a large number of environmental antigens immediately after birth could have significant detrimental consequences in terms of an overwhelming immune response.

3. The presence of key commensal microbes in the fetus may promote colonization resistance, that is, prevention of the establishment of more pathogenic microbes.
4. The technological developments (next-generation sequencing and bioinformatics) that have led to major improvements in our ability to identify and characterize microbes that are often very difficult to culture provide evidence of microbial component presence, but not actually the presence of stable colonization by live microbes in various niches.
5. Studies utilizing both culture and nonculture techniques have shown that upper female reproductive tract is not sterile [14]. The placenta implants into uterine tissue that is highly likely to be exposed to the upper vaginal and fallopian tube microbiota, which differ markedly from those found in the lower vagina [14].
6. Although it has been stated that contamination and the "kitome" are confounding factors that mitigate against an in utero microbiome [31,32,65,66], a large number of the published data do, in fact [28,67],

take contamination into account, describe a low biomass, and *do not claim colonization*. Some of these studies demonstrate visualization of intact microbes in placental tissue, form both term and preterm gestations distinct from contamination controls [27,68]. Other studies have demonstrated the ability to culture microbes from preterm and term placenta [69–72].

7. Studies in animals and humans do demonstrate microbial presence in various fetal sites including the sheep brain [73], the mouse placenta and fetal intestine (where overlap between taxa was found) [74], and human fetal intestine [46].
8. Even if microbes are not directly present in the womb, their colonization in distal sites in the mother may still play a major role in fetal immune development. This is supported by studies of antenatal monocolonization of pregnant germ-free mice, which reprogrammed intestinal transcriptional profiles of the offspring, increasing expression of genes encoding genes for epithelial antibacterial peptides as well as increasing various innate lymphoid cells in the offspring for months after birth [75]. This is further supported by studies showing bacterial DNA and bacterially derived short chain fatty acids are present in fetal gut and amniotic fluid, and they have the potential to influence the developing immune system [61].
9. Sample binning remains imperfect, bioinformatic pipeline development remains a work in progress, and sharing taxa with controls does not necessarily imply contamination.
10. The ability to generate germ-free mice by delivering fetuses via cesarean section and raising the offspring in sterile environments has been used to refute the idea that the placenta contains microbes [33]. Germ-free animals are difficult to generate and exhibit very different phenotypes than conventionally raised animals from the time of birth, suggesting that microbial stimulation, even if not by direct in utero contact, is still required for normal development.
11. The fetus has a functional and developing immune system at the time of birth, which includes tolerance, plasma cells capable of immunoglobulin production (including IgM), and microglia in the brain.

Why is an in utero microbiome important?

It is clear that a large number of critically important metabolites such as serotonin, vitamins, and short chain fatty acids rely on their production by microbes. Interaction of microbial components with the host immune system is critical for immunologic maturation. These all play a major role in the host's subsequent lifetime health. Next, some of these aspects will be briefly summarized.

Mode of delivery and health

In the early 21st century, several studies reported that childhood or infantile atopic diseases were related to the method of delivery of those children. These studies showed that elective cesarean deliveries were associated with an increased risk of allergic diseases in childhood. In 2004 Negele et al. [76] reported a study performed in Germany, which assessed infants over their first 2 years and specifically looked at mode of delivery and development of allergies. They found that cesarean section was associated with a higher incidence of all tested atopic diseases and diagnoses within this 2-year follow-up period [76]. These results led the group to question how cesarean section affects the intestinal microbiome around the time of delivery, given the importance of intestinal lymph tissue in the development of the infant immune system. Studies have shown differences in the colonization of the newborn depending on mode of delivery and have related this to previous epidemiologic studies showing differences in health outcomes [41].

Gregory et al. analyzed how mode of delivery affected *Bacteroides* colonization in preterm infants [77]. *Bacteroides* is an anaerobic genus of bacteria that colonizes most gastrointestinal tracts [78]. It is typically acquired through vaginal delivery and Gregory evaluated how cesarean delivery affected time significant colonization [78]. Gregory found that delivery via the vaginal canal produced a significantly greater colonization as well as greater populations of *Bacteroides* as time elapsed. Cesarean deliveries by contrast had much lower concentrations of Bacteroides at delivery and showed a much slower growth. Given the immune modulating effects that *Bacteroides* can have, this delay may result in an increased risk for infections for cesarean delivered neonates.

If there is a prenatal microbiome, then "seeding," in fact, likely occurs prior to delivery and the need for providing vaginally derived microbes to cesarean section delivered infants may be minimal. Thus vaginal microbe "seeding," even though vaginally derived microbes may be transmitted to the newborn, as described in recent studies [79], may not be as critical to health as the prenatal exposure. In fact, large studies now suggest that a considerable amount of the difference in microbial development postnatally when comparing mode of delivery is based on confounding factors such as antibiotic use, feeding, and length of time in the hospital allowing acquisition of hospital microbes [80,81].

Development of the infant immune system

Multiple studies have produced evidence that in utero exposure to bacteria affects the fetal/neonatal immune system, specifically with regard to autoimmune disease, metabolic disease, allergies, and gastrointestinal disease

[82,83]. As these diseases are usually diagnosed later in life and have lifelong implications, finding evidence that alterations in the fetal gastrointestinal microbiome leading to such diseases makes research into this area vital.

Fetal and neonatal immune systems are in the early stages of development and take many months to develop fully. It is for this reason that vaccinations are repeated so many times within the first few years after birth, as the immune system needs boosters to help aide in recognition of the vaccine components. Maternal contributions, both in utero and via breast milk after delivery, help to protect the infant as this immune system continues to develop by passively transferring immunoglobulins to the baby. The gastrointestinal tract is known to have lymph tissue scattered throughout it (Peyer's patches) and has been theorized as the source of initial exposure to environmental microbes during early development. The fetal and neonatal intestinal microbiome then would be the first exposure of this developing immune system and any perturbations within this microbiome could have far reaching effects on the body's long-term immune response. Delays in feeding as seen with prematurity and in sick full-term infants would lead to decreased use of the gastrointestinal system and delay in maturity of Peyer's patches. Antibiotics, mode of delivery, type of feeding, hospitalization, and other environmental factors could alter the microbiome after delivery and therein change what bacteria the infants are exposed to.

Metabolism

In 2014 Cox and Blaser reported metabolic concerns after disruption of the early microbiome with relation to early administration of antibiotics or cesarean deliveries [84]. This disruption led to an increased risk of obesity within childhood, increase in fat mass, altered metabolic hormones, and an altered microbial composition within these children. The changes within the microbiome in particular seemed to have lasting metabolic changes. Bacterial composition affected by cesarean delivery was left with a more limited spectrum than vaginally delivered counterparts. Those affected by antibiotics could lose a portion of the enteric microbiome and develop new variants that are able to survive the antibiotics. This study supports the idea that these changes, even when limited to the first few years of life, can lead to differences in adiposity as far reaching as early to mid-adulthood. This appears to correlate with ideas proposed in adult medical literature about associations between metabolic syndrome and the adult microbiome. Further research into the neonatal and early infant microbiome could then lead to advances in prevention of such metabolic disorders.

Conclusion

From this review, there should be no question that microbes play a role in fetal and subsequent development. However, the exact proximity of live beneficial microbes to the placenta and fetus, whether they are embedded in the placenta or fetus in their own niche (true colonization), whether they need to be viable and functioning in immediate proximity to the developing tissues of interest, all are issues that require further scrutiny.

References

[1] Willyard C. Could baby's first bacteria take root before birth? Nature 2018;553:264–6.

[2] Tissier H. Recherches sur la flore intestinale des nourrissons (etal normal et pahthologique) [M.D. thesis]. Paris; 1900.

[3] Hugenholtz P. Exploring prokaryotic diversity in the genomic era. Genome Biol 2002;3 Reviews0003.

[4] Staley JT, Konopka A. Measurement of in situ activities of nonphotosynthetic microorganisms in aquatic and terrestrial habitats. Annu Rev Microbiol 1985;39:321–46.

[5] Woese CR, Kandler O, Wheelis ML. Towards a natural system of organisms: proposal for the domains Archaea, Bacteria, and Eucarya. Proc Natl Acad Sci U S A 1990;87:4576–9.

[6] NIH Human Microbiome Portfolio Analysis Team. A review of 10 years of human microbiome research activities at the US National Institutes of Health, Fiscal Years 2007–2016. Microbiome 2019;7:31.

[7] Knight R, Vrbanac A, Taylor BC, et al. Best practices for analysing microbiomes. Nat Rev Microbiol 2018;16:410–22.

[8] Doderlein A. Das scheidensekret und seine bedeutung fur puerperalfieber. Zbl Bakteriol 1892;11:699.

[9] Witkin SS, Linhares IM, Giraldo P. Bacterial flora of the female genital tract: function and immune regulation. Best Pract Res Clin Obstet Gynaecol 2007;21:347–54.

[10] Goldenberg RL, Hauth JC, Andrews WW. Intrauterine infection and preterm delivery. N Engl J Med 2000;342:1500–7.

[11] Goldenberg RL, Culhane JF, Iams JD, Romero R. Epidemiology and causes of preterm birth. Lancet 2008;371:75–84.

[12] Smith SB, Ravel J. The vaginal microbiota, host defence and reproductive physiology. J Physiol 2017;595:451–63.

[13] Ravel J, Gajer P, Abdo Z, et al. Vaginal microbiome of reproductive-age women. Proc Natl Acad Sci U S A 2011;108:4680–7.

[14] Chen C, Song X, Wei W, et al. The microbiota continuum along the female reproductive tract and its relation to uterine-related diseases. Nat Commun 2017;8:875.

[15] Harris JW, Brown JH. The bacterial content of the uterus at cesarean section. Am J Obstet Gynecol 1927;13:133–43.

[16] Harwick HJ, Iuppa JB, Fekety FR. Microorganisms and amniotic fluid. Obstet Gynecol 1969;33:256–9.

[17] Bobitt JR, Ledger WJ. Unrecognized amnionitis and prematurity: a preliminary report. J Reprod Med 1977;19:8–12.

[18] Romero R, Espinoza J, Gonçalves LF, Kusanovic JP, Friel L, Hassan S. The role of inflammation and infection in preterm birth. Semin Reprod Med 2007;25:21–39.

[19] Dammann O, Kuban KC, Leviton A. Perinatal infection, fetal inflammatory response, white matter damage, and cognitive limitations in children born preterm. Ment Retard Dev Disabil Res Rev 2002;8:46–50.

[20] Dammann O, Leviton A. Inflammation, brain damage and visual dysfunction in preterm infants. Semin Fetal Neonatal Med 2006;11:363–8.

[21] DiGiulio DB, Gervasi MT, Romero R, et al. Microbial invasion of the amniotic cavity in pregnancies with small-for-gestational-age fetuses. J Perinat Med 2010;38.

[22] DiGiulio DB. Diversity of microbes in amniotic fluid. Semin Fetal Neonatal Med 2012;17:2–11.

[23] Aquino TI, Zhang J, Kraus FT, Knefel R, Taff T. Subchorionic fibrin cultures for bacteriologic study of the placenta. Am J Clin Pathol 1984;81.

[24] Arora N, Sadovsky Y, Dermody TS, Coyne CB. Microbial vertical transmission during human pregnancy. Cell Host Microbe 2017;21:561–7.

[25] Kovalovszki L, Villanyi Z, Pataki I, Veszelowvsky I, Nagy ZB. Isolation of aerobic bacteria from the placenta. Acta paediatrica Academiae Scientiarum Hungaricae 1982;23:357–60.

[26] Hillier SL, Martius J, Krohn M, Kiviat N, Holmes KK, Eschenbach DA. A case-control study of chorioamnionic infection and histologic chorioamnionitis in prematurity. N Engl J Med 1988;319:972–8.

[27] Stout MJ, Conlon B, Landeau M, et al. Identification of intracellular bacteria in the basal plate of the human placenta in term and preterm gestations. Am J Obstet Gynecol 2013;208.

[28] Aagaard K, Ma J, Antony KM, Ganu R, Petrosino J, Versalovic J. The placenta harbors a unique microbiome. Sci Transl Med 2014;6:237–65.

[29] de Goffau MC, Lager S, Salter SJ, et al. Recognizing the reagent microbiome. Nat Microbiol 2018;3:851–3.

[30] The Lancet Infectious Diseases. Microbiome studies and "blue whales in the Himalayas". Lancet Infect Dis 2018;18:925.

[31] Lauder AP, Roche AM, Sherrill-Mix S, et al. Comparison of placenta samples with contamination controls does not provide evidence for a distinct placenta microbiota. Microbiome 2016;4:29.

[32] de Goffau MC, Lager S, Sovio U, et al. Human placenta has no microbiome but can contain potential pathogens. Nature 2019;572:329–34.

[33] Perez-Muñoz ME, Arrieta M-C, Ramer-Tait AE, Walter J. A critical assessment of the "sterile womb" and "in utero colonization" hypotheses: implications for research on the pioneer infant microbiome. Microbiome 2017;5:48.

[34] Chen HJ, Gur TL. Intrauterine microbiota: missing, or the missing link? Trends Neurosci 2019;42:402–13.

[35] Zhu L, Luo F, Hu W, et al. Bacterial communities in the womb during healthy pregnancy. Front Microbiol 2018;9:2163.

[36] Wright TE. Biochemical screening for in utero drug exposure. Drug Metab Lett 2015;9:65–71.

[37] Meetze WH, Palazzolo VL, Bowling D, Behnke M, Burchfield DJ, Neu J. Meconium passage in very-low-birth-weight infants. JPEN J Parenter Enteral Nutr 1993;17:537–40.

[38] Lin JK, Huang TS, Luh KT. Penicillinase-type of beta-lactamase responsible for the ampicillin resistance in *Escherichia coli* NTUH 9501-1. Taiwan I Hsueh Hui Tsa Chih 1989;88:1–7.

[39] Hau IC, O'Toole E, Hall IP. Bacterial flora of first specimens of meconium passed by fifty newborn infants. Am J Dis Child 1960;1934:47.
[40] Biasucci G, Benenati B, Morelli L, Bessi E, Boehm G. Cesarean delivery may affect the early biodiversity of intestinal bacteria. J Nutr 2008;138.
[41] Dominguez-Bello MG, Costello EK, Contreras M, et al. Delivery mode shapes the acquisition and structure of the initial microbiota across multiple body habitats in newborns. Proc Natl Acad Sci U S A 2010;107:11971–5.
[42] Jimenez E, Marin ML, Martin R, et al. Is meconium from healthy newborns actually sterile? Res Microbiol 2008;159:187–93.
[43] Mshvildadze M, Neu J, Shuster J, Theriaque D, Li N, Mai V. Intestinal microbial ecology in premature infants assessed with non-culture-based techniques. J Pediatr 2010;156:20–5.
[44] Ardissone AN, de la Cruz DM, Davis-Richardson AG, et al. Meconium microbiome analysis identifies bacteria correlated with premature birth. PLoS One 2014;9:e90784.
[45] Hansen R, Scott KP, Khan S, et al. First-pass meconium samples from healthy term vaginally-delivered neonates: an analysis of the microbiota. PLoS One 2015;10.
[46] Younge N, McCann JR, Ballard J, et al. Fetal exposure to the maternal microbiota in humans and mice. JCI Insight 2019;4.
[47] Gomez-Arango LF, Barrett HL, McIntyre HD, Callaway LK, Morrison M, Nitert MD. Contributions of the maternal oral and gut microbiome to placental microbial colonization in overweight and obese pregnant women. Sci Rep 2017;7:2860.
[48] Koren O, Goodrich JK, Cullender TC, et al. Host remodeling of the gut microbiome and metabolic changes during pregnancy. Cell 2012;150:470–80.
[49] Zhang D, Huang Y, Ye D. Intestinal dysbiosis: an emerging cause of pregnancy complications? Med Hypotheses 2015;84:223–6.
[50] Jimenez E, Fernandez L, Marin ML, et al. Isolation of commensal bacteria from umbilical cord blood of healthy neonates born by cesarean section. Curr Microbiol 2005;51:270–4.
[51] Rautava S, Collado MC, Salminen S, Isolauri E. Probiotics modulate host-microbe interaction in the placenta and fetal gut: a randomized, double-blind, placebo-controlled trial. Neonatology 2012;102:178–84.
[52] Potgieter M, Bester J, Kell DB, Pretorius E. The dormant blood microbiome in chronic, inflammatory diseases. FEMS Microbiol Rev 2015;39:567–91.
[53] Thum C, Cookson AL, Otter DE, et al. Can nutritional modulation of maternal intestinal microbiota influence the development of the infant gastrointestinal tract? J Nutr 2012;142:1921–8.
[54] Bailey MT, Lubach GR, Coe CL. Prenatal stress alters bacterial colonization of the gut in infant monkeys. J Pediatr Gastroenterol Nutr 2004;38:414–21.
[55] Mayer EA, Knight R, Mazmanian SK, Cryan JF, Tillisch K. Gut microbes and the brain: paradigm shift in neuroscience. J Neurosci 2014;34:15490–6.
[56] Sudo N, Chida Y, Aiba Y, et al. Postnatal microbial colonization programs the hypothalamic-pituitary-adrenal system for stress response in mice. J Physiol (Lond) 2004;558:263–75. Available from: https://doi.org/10.1113/jphysiol.2004.063388.
[57] Messaoudi M, Lalonde R, Violle N, et al. Assessment of psychotropic-like properties of a probiotic formulation (*Lactobacillus helveticus* R0052 and *Bifidobacterium longum* R0175) in rats and human subjects. Br J Nutr 2011;105:755–64.
[58] Shapiro GD, Fraser WD, Frasch MG, Seguin JR. Psychosocial stress in pregnancy and preterm birth: associations and mechanisms. J Perinat Med 2013;41:631–45.

[59] Gur TL, Worly BL, Bailey MT. Stress and the commensal microbiota: importance in parturition and infant neurodevelopment. Front Psychiatry 2015;6:5.
[60] Moles L, Gomez M, Heilig H, et al. Bacterial diversity in meconium of preterm neonates and evolution of their fecal microbiota during the first month of life. PLoS One 2013;8: e66986.
[61] Stinson LF, Boyce MC, Payne MS, Keelan JA. The not-so-sterile womb: evidence that the human fetus is exposed to bacteria prior to birth. Front Microbiol 2019;10:1124.
[62] Nanthakumar NN, Fusunyan RD, Sanderson I, Walker WA. Inflammation in the developing human intestine: a possible pathophysiologic contribution to necrotizing enterocolitis. Proc Natl Acad Sci U S A 2000;97:6043–8.
[63] Nanthakumar N, Meng D, Goldstein AM, et al. The mechanism of excessive intestinal inflammation in necrotizing enterocolitis: an immature innate immune response. PLoS One 2011;6:e17776.
[64] Funkhouser LJ, Bordenstein SR. Mom knows best: the universality of maternal microbial transmission. PLoS Biol 2013;11:e1001631.
[65] Theis KR, Romero R, Winters AD, et al. Does the human placenta delivered at term have a microbiota? Results of cultivation, quantitative real-time PCR, 16S rRNA gene sequencing, and metagenomics. Am J Obstet Gynecol 2019;220 267.e1–.e39.
[66] Leiby JS, McCormick K, Sherrill-Mix S, et al. Lack of detection of a human placenta microbiome in samples from preterm and term deliveries. Microbiome 2018;6:196.
[67] Antony KM, Ma J, Mitchell KB, Racusin DA, Versalovic J, Aagaard K. The preterm placental microbiome varies in association with excess maternal gestational weight gain. Am J Obstet Gynecol 2015;212.
[68] Seferovic MD, Pace RM, Carroll M, et al. Visualization of microbes by 16S in situ hybridization in term and preterm placentas without intraamniotic infection. Am J Obstet Gynecol 2019;221 146.e1–.e23.
[69] Onderdonk AB, Hecht JL, McElrath TF, Delaney ML, Allred EN, Leviton A. Colonization of second-trimester placenta parenchyma. Am J Obstet Gynecol 2008;199 52 e1–e10.
[70] Fichorova RN, Onderdonk AB, Yamamoto H, et al. Maternal microbe-specific modulation of inflammatory response in extremely low-gestational-age newborns. mBio 2011;2 e00280-10.
[71] Collado MC, Rautava S, Aakko J, Isolauri E, Salminen S. Human gut colonisation may be initiated in utero by distinct microbial communities in the placenta and amniotic fluid. Sci Rep 2016;6.
[72] Parnell LA, Briggs CM, Cao B, Delannoy-Bruno O, Schrieffer AE, Mysorekar IU. Microbial communities in placentas from term normal pregnancy exhibit spatially variable profiles. Sci Rep 2017;7:11200.
[73] Zarate MA, Rodriguez MD, Chang EI, et al. Post-hypoxia invasion of the fetal brain by multi-drug Staphylococcus. Sci Rep 2017;23(1):6458. https://doi:10.1038/s41598-017-06789-6.
[74] Martinez 2nd KA, Romano-Keeler J, Zackular JP, et al. Bacterial DNA is present in the fetal intestine and overlaps with that in the placenta in mice. PLoS One 2018;13: e0197439.
[75] Gomez De Aguero M, Ganal-Vonarburg SC, Fuhrer T, et al. The maternal microbiota drives early postnatal innate immune development. Science 2016;351.
[76] Negele K, Heinrich J, Borte M, et al. Mode of delivery and development of atopic disease during the first 2 years of life. Pediatr Allergy Immunol 2004;15:48–54.

[77] Gregory KE, LaPlante RD, Shan G, Kumar DV, Gregas M. Mode of birth influences preterm infant intestinal colonization with bacteroides over the early neonatal period. Adv Neonatal Care 2015;15.
[78] Wexler HM. Bacteroides: the good, the bad, and the nitty-gritty. Clin Microbiol Rev 2007;20:593–621.
[79] Dominguez-Bello MG, Jesus-Laboy KM, Shen N, et al. Partial restoration of the microbiota of cesarean-born infants via vaginal microbial transfer. Nat Med 2016;22.
[80] Chu DM, Ma J, Prince AL, Antony KM, Seferovic MD, Aagaard KM. Maturation of the infant microbiome community structure and function across multiple body sites and in relation to mode of delivery. Nat Med 2017;23:314–26.
[81] Reyman M, van Houten MA, van Baarle D, et al. Impact of delivery mode-associated gut microbiota dynamics on health in the first year of life. Nat Commun 2019;10:4997.
[82] Stinson LF, Payne MS, Keelan JA. Planting the seed: origins, composition, and postnatal health significance of the fetal gastrointestinal microbiota. Crit Rev Microbiol 2017;43:352–69.
[83] Van Belkum M, Mendoza Alvarez L, Neu J. Preterm neonatal immunology at the intestinal interface. Cell Mol Life Sci 2020;77(7):1209–27. https://doi:10.1007/s00018-019-03316-w.
[84] Cox LM, Blaser MJ. Pathways in microbe-induced obesity. Cell Metab 2013;17:883–94.

Chapter 2

Microbiome establishment and maturation: early life environmental factors

Alyson L. Yee[1], Promi Das[2], Mariana C. Salas Garcia[2], Mary M. Buschmann[2] and Jack A. Gilbert[2]
[1]Interdisciplinary Scientist Training Program, University of Chicago, Chicago, IL, United States, [2]Department of Pediatrics, University of California San Diego, La Jolla, CA, United States

Introduction

The human microbiome is the sum of all microorganisms, viruses, and their genes present in the human body. This microbial ecosystem presents a complex ecology that requires the application of ecological theory to account for variance over time within an individual and to determine the differences between individuals [1]. We use ecological theory to examine how the microbial ecosystem varies between and within people and how this variance associates with disease through the application of microbiome wide association studies (MWAS) [2]. Although the various constituents and functions of the microbiome have been found to have associative and causative influence on disease [3], how the microbiome is altered by intrinsic and extrinsic forces remains less understood. As a field, we do not yet understand what drives the differences in the microbiome between various people and within people over time. When performing microbiome studies in humans, we have to account for the myriad factors that have been shown to cause changes in the microbial community, including diet, exercise, age, medications, and even geography. While examining these phenomena in infants we also have to account for the fact that the microbial community is undergoing rapid ecological succession, which compounds the observed variance in community structure both within and between subjects.

The microbiome tends to stabilize around a "adult-like" community structure and function within the first 2–3 years of life. The period of development during this phase, when the microbiome is undergoing dynamic

The Developing Microbiome. DOI: https://doi.org/10.1016/B978-0-12-820602-7.00002-7

equilibrium, could be vital for training a child's immune system, allowing for appropriate digestion and nutrient absorption, as well as protecting the child from disease. Therefore identifying the factors that shape the colonization, succession, and maturation of the microbiome and determining how this trend varies between children and how it associates with disease burden will have implications for treatment of disease. Therefore characterizing the factors that drive the successional dynamics of the infant microbiome should be a research priority. Various studies have shown that the pioneer gut microbiome is heavily influenced by maternal and perinatal factors, and it may affect subsequent ecological succession, implying that initial exposure is an important determinant of the later composition of the gut microbiome. Here, we consider the effects of maternal microbial exposures (prenatal); peripartum factors (delivery mode, antibiotics, hospitalization); nutrition (breastfeeding vs formula, the introduction of solid food); and environment (family, pets, indoor microbiome) on the microbial colonization of infants. All of these exposures have been shown to have a lasting effect on the microbiome of children with potential associations with their health.

Maternal exposures

Microbial species present in the maternal and infant gut overlap in the first year of life [4,5]. Many microbial taxa are heritable [6], although the environment seems to predominate over genetics in shaping the microbiome [7,8]. Still, maternal influences are important as both genetic and environmental determinants of the infant microbiota. While intrauterine microbial exposure is a matter of debate, and more thoroughly reviewed in Chapter 1, prenatal changes in the maternal microbiome have been considered in relation to their effects on the infant microbiome. For example, antibiotic use during pregnancy has been associated with lower diversity and abundance of *Lactobacillus* and *Bifidobacterium* in the neonate's stool [9–11]. Even without definitive evidence of bacteria crossing the placenta, the fetus is certainly exposed to metabolites and compounds (including DNA) originating from the maternal microbiota, which may affect development [12,13]. Metabolic events that take place in fetal life may have lasting implications for future health. For example, adult adiposity, osteoporosis, depression, and other diseases have been linked with poor maternal nutrition [14]. These epidemiological studies, along with the sequencing studies associating maternal factors with infant microbiota, underscore the importance of the intrauterine environment for immune development and long-term health.

The idea that maternal health can affect offspring health via modulation of the microbiome is receiving increasing attention. A study of maternal weight showed that infants born from women with normal weight gain during pregnancy harbored higher concentrations of *Bifidobacterium* than those whose mothers gained excess weight [15]. The same pattern was observed in the mother's gut microbiota, with higher weight linked to lower numbers of *Bifidobacterium* [15]. A nonhuman primate study showed that maternal high-fat diet during pregnancy and breastfeeding altered the offspring microbiome [16]. Similar to the primate model, a population-based human cohort showed that maternal high-fat diet was associated with distinctive signatures in the neonatal gut microbiome, independent of maternal body mass index [17]. A recent study in human mother–infant dyads identified maternal fruit intake during pregnancy as a driver of neonatal microbiome composition in vaginally delivered, but not cesarean-delivered, children [18]. The authors observed differences in both overall infant gut microbial community structure as well as abundances of specific microbes that related to maternal dietary factors in a delivery mode-dependent pattern [18]. Other studies have observed that mothers with eczema had infants with decreased meconium diversity and mothers with gestational diabetes passed a distinct microbiota to their infants [9,19,20].

Novel metagenomics approaches have been used to characterize strain-level variation in the microbiome. This allows for tracking of specific single nucleotide polymorphisms and provides evidence of transmission patterns [5,21]. Maternal skin and vaginal strains were shown to colonize transiently, and infants continued to acquire microbes from maternal sources after birth [21]. Compared with bacteria from other sources, maternal gut strains were the most persistent in the infant gut, implying ecological adaptation to the gut niche [21]. Together, the consensus in the literature is that the mother is the infant's most important source of microbial colonization. Maternal influence begins during gestation and continues to shape the infant's development postnatally, as we will discuss in the following sections.

Delivery mode

While the extent of in utero microbial exposures remains controversial (and is more fully considered in Chapter 1), the majority of bacterial colonization occurs during and immediately following birth. Colonization is thought to be rapid; experiments in germ-free rodents have shown that bacteria are detectable within hours after initial exposure to conventional housing and that bacterial counts are equivalent to those in conventional mice within 24 hours [22]. Therefore birth represents a massive colonization event, and mode of delivery has been consistently shown to affect initial gut microbiome composition.

In vaginal delivery, the infant comes into contact with maternal vaginal and enteric contents [23,24]. Vaginally delivered infants are colonized with bacteria from the taxa *Lactobacillus*, *Prevotella*, *Bacteroides*, *Escherichia/Shigella*, and *Bifidobacterium*, which are also present in vaginal and fecal samples of adult mothers [25]. However, in the United States, >30% of births are now cesarean deliveries, so understanding the effects of mode of delivery on the initial assembly of the microbiome may have wide-ranging impacts [24,26]. Prelabor cesarean-delivered infants, in contrast, are dominated by human skin-associated and oral bacteria, including *Staphylococcus*, *Streptococcus*, *Corynebacterium*, *Veillonella*, and *Propionibacterium* [4,25]. Despite high variability between infants, studies have consistently detected the effect of delivery mode on the relative abundances of several bacterial taxa; cesarean delivery is associated with lower *Escherichia–Shigella* and an absence of *Bacteroides* [27]. Postnatal colonization of the neonatal gut by *Bacteroides* and *Bifidobacterium* spp. seems to be delayed in cesarean-delivered infants [9,26,28,29]. Some studies have found that the effect of delivery mode persists longer than differences by gestational age [30], whereas others have found that the effect of mode of delivery is eclipsed by that of preterm birth [31–34]. Specific concerns for preterm delivery will be more fully discussed in Chapter 4, but distinct patterns of colonization by mode of delivery are widely accepted for healthy, term infants.

Differences in the microbiome community composition between vaginally and cesarean-delivered infants gradually decrease, but distinct gut microbial profiles associated with delivery mode may persist up to 4 or more years of age [26,35]. In contrast, other groups have argued that there is no discernible effect past the immediate neonatal period [8,12,36]. These classic studies showed strong associations between mode of delivery and differences in microbial composition. Recent advances in high throughput sequencing and higher resolution strain analysis have allowed for higher resolution culture-independent study of transmission of microbes. On a strain-resolved level too, maternal seeding is affected by birth mode, with strikingly different strain transmission and persistence patterns in vaginally versus cesarean section-delivered infants [37]. In particular, Korpela et al. did not observe sharing of a single strain between cesarean-delivered infants and their mothers, implicating vaginal birth as the main transmission route [37]. They concluded that cesarean-delivered infants fail to receive maternal strains at birth and instead show a high degree of strain flux in early life, comparable to the flux observed in nontransmitted species in vaginally born infants [37].

Antibiotics

Antibiotics reduce microbial load and diversity in adults and animal models [38,39], and such treatments are among the most widely used therapies in pediatric populations [40]. Although the microbiota is resilient to perturbation by antibiotics, returning to stable equilibrium states after treatment [41], antibiotic usage in early life or for long periods of time may affect the gut microbiota. Indeed, antibiotic-related disruption of the resident microbiota may leave hosts more vulnerable to enteric pathogens such as *Clostridium difficile* [42]. Frequent or continuous use of postnatal antibiotics in early life has been correlated with increased risk for many conditions, including asthma, inflammatory bowel disease, type 2 diabetes, and food allergy [43–48]. In general, bactericidal drugs decrease the overall diversity of the infants' gut microbiota and appear to favor the growth of drug-resistant microbes [49–51]. Fouhy et al. showed that infants exposed to ampicillin and gentamicin shortly after birth had greater proportions of Proteobacteria, Actinobacteria, and *Lactobacillus* than the unexposed children for up to 4 weeks after concluding treatment [52]. Bokulich et al. examined the time course of microbiota maturation using regression models to predict age as a function of microbial composition [53]. They found that children exposed to antibiotics had delayed microbiota maturation, particularly during 6–12 months of age, due to depletion of specific taxa such as Enterobacteriaceae, Lachnospiraceae, and Erysipelotrichaceae [53].

Yassour et al. leveraged metagenomics to identify strain-level differences in longitudinal samples from the same children before and after antibiotic treatment. They concluded that children who received antibiotics had less gut microbial alpha diversity, though this was modest and was only detectable a year after birth [54]. However, not all bacterial species were affected the same way, with more significant effects of antibiotics on strain similarity shown in those species that usually only had a single colonizer [54]. Furthermore, children who received antibiotics had significantly less stable microbial communities surrounding treatment [54]. In addition, antibiotic resistance genes on microbial chromosomes peaked in abundance after antibiotic treatment and then declined, while antibiotic resistance genes carried on mobile elements tended to persist after the end of therapy [54].

In a cohort of seven infants who received antibiotics in a neonatal intensive care unit (NICU), antibiotics given at birth were found to alter the microbiome, disrupting the abundance of microbial genes involved in four metabolic pathways, whereas courses given later did not [55]. Comparing longitudinal samples before, during, or following antibiotics, Bender et al. discovered that there was more similarity between samples collected while off antibiotics than in the transition from off-to-on antibiotics [55]. A study of a preschool population showed clear differences in the microbiota composition in the antibiotic-treated children as compared with those who were not exposed to antibiotics for >2 years [56]. They determined that there were differential effects by class of antibiotic, with macrolides affecting the abundance of *Bifidobacterium* and *Bacteroides*, as well as macrolide resistance, for about 12 months, and the abundances of *Collinsella*, *Lactobacillus*, and *Anaerostipes*, as well as the total richness and maturity of the microbiota, remaining reduced for up to 2 years after a macrolide course. They concluded that a full recovery of the microbiota from an antibiotic course might take longer than the usual interval between courses, in contrast to some studies of adult populations [56].

Interestingly, the microbiota at different body sites appears to respond differently to antibiotic perturbation, with the salivary microbiome recovering more quickly than the fecal microbiome to disruption with clindamycin, ciprofloxacin, amoxicillin, and minocycline [57]. Consistent with other reports, the fecal microbiome was depleted in butyrate-producing species, enriched for antibiotic resistance genes, and these changes persisted for months following cessation of antibiotic exposure [57]. In animal models, even a single course of early life antibiotic exposure was sufficient to perturb the gut microbiota network topology well past the cessation of treatment [58]. This could explain why decreased diversity is stable with pulsed therapy or following transfer of an antibiotic-perturbed microbiota to germ-free recipients [38,58]. In a mouse model of antibiotic exposure before weaning, mice exposed to low-dose penicillin before weaning had an altered gut microbiota compared to controls, including a decrease in *Lactobacillus* and *Allobaculum*, but the community composition recovered to resemble the control communities after the antibiotics were removed [59]. However, despite recovery of the microbiota by 16S rRNA amplicon sequencing proportion, mice that were exposed to antibiotics had increased mass even 26 weeks after the antibiotic treatment was stopped [59]. This increased mass was only observed if the antibiotics were administered before weaning, lending support to the observation that early life is a critical window for immune and metabolic development [59,60]. Taken together, these notable studies imply that antibiotic use represents a perturbation that affects the ecology of the infant gut.

Nutrition

Nutrition has a major impact on microbiota composition. In infants, the main dietary modalities that have been studied are exclusive breastfeeding and formula feeding. Breastfeeding is widely accepted as beneficial to both mothers and infants, as it is a nutritious food and its composition varies over the course of lactation with the infant's changing nutritional requirements [61]. In addition to containing the appropriate nutrients, breast milk is a source of prebiotic oligosaccharides that are considerably different from the carbohydrates present in cow's milk [61]. Furthermore, breast milk contains a diverse population of bacteria that are hypothesized to seed the infant's gut and that have evolved to metabolize these oligosaccharides specific to human breast milk [62]. Culture-based studies have tracked the transmission of individual members of the microbial communities, providing direct evidence of breastfeeding as a source of microbial colonization in infants [63]. Recently, sequencing-based approaches have lent further support to the transmission of bacteria from mother to infant via feeding. Overall, the literature points to significant taxonomic and functional differences in microbial composition between infants who receive breast milk and those who receive formula [4,27,64,65]. Breastfed infants have reduced richness and alpha diversity compared with those who receive formula, but this richness negatively correlates with *Bifidobacterium* abundance, suggesting that breast milk selects for *Bifidobacterium* predominance [27]. In addition to *Bifidobacterium*, breastfeeding has been associated with bacterial species frequently used as probiotics, such as *Lactobacillus* spp. [4,62,65].

In a population of 98 full-term infants in Sweden, clear differences were observed between infants who received exclusive breastfeeding and exclusive formula feeding [4]. Infants who were exclusively breastfed had increased levels of taxa including *L. johnsonii*, *L. gasseri*, *L. paracasei*, *L. casei*, and *B. longum*, whereas 4-month-old formula-fed infants had elevated levels of *Clostridium difficile*, *Granulicatella adiacens*, *Citrobacter* spp., *Enterobacter cloacae*, *Bilophila wadsworthia* [4]. The authors further probed functional gene differences between the feeding modes and found that although feeding pattern played a small role in explaining the variance in functional genes, formula-fed infants were enriched in functions found in the adult microbiome, such as bile acid biosynthesis and methanogenesis. Meanwhile, the microbiome of infants that were exclusively breastfed had higher levels of KEGG orthology group modules involved in oxidative phosphorylation and synthesis of B vitamins [4].

In The Environmental Determinants of Diabetes in the Young (TEDDY) study, surveying longitudinal microbiome development in 903 children, receipt of breast milk explained the greatest amount of variance in the microbiome structure from months 3 to 14 [65]. Breastfeeding was significantly associated with 121 different bacterial species, with *Bifidobacterium* higher in children who received breast milk either exclusively or together with formula milk or solids [65]. Notably, TEDDY study identified trends in species that have been found in breast milk (*Bifidobacterium* spp. and *Lactobacillus* spp.) as well as species that colonize areolar skin (*Staphylococcus* spp.), implying that these species can be directly transferred from the mother to infant [65]. Furthermore, *Bifidobacterium* species that did not correlate with breastfeeding were shown to vary in the abundance of genes implicated in human milk oligosaccharide metabolism before and after cessation of breastfeeding [66]. These strains had distinct profiles of sugar utilization, underscoring the role of nutrient availability in shaping the gut microbial community [66].

Pannaraj et al. analyzed samples from 107 mother–infant pairs to determine how bacteria from maternal breast milk and areolar skin transferred to infant guts. They found breastfed infants received 27.7% of their gut bacteria from breast milk and 10.4% from areolar skin during the first month of life [62]. They also identified associations between bacterial diversity and the proportion of daily breast milk intake in a dose-dependent manner, even after introduction of solid foods [62]. Therefore, maternal factors that affect breast milk nutritional content and microbiome composition can reasonably be expected to have downstream effects on breastfed infants. Indeed, bacterial genes such as fucosidases and ureases, implicated in using glycans and amino acids from maternal breast milk, vary between Amerindian, Malawian, and US infant microbiomes, correlating with the differences in diet between these populations [64].

Studies that have examined the role of breastfeeding have varied in terms of amount and duration of breastfeeding and have been confounded by other peripartum factors. However, in a meta-analysis of seven studies, infants who were exclusively breastfed consistently showed lower gut bacterial diversity, lower microbiota age, lower relative abundances of Bacteroidetes and Firmicutes, and lower abundance of pathways related to microbial carbohydrate metabolism compared with children who received some formula [67]. Furthermore, infants who received formula tended to have lower abundances of pathways related to lipid metabolism, vitamin metabolism, and detoxification [67].

As children grow, the cessation of breastfeeding is an important transition, with data suggesting the main driver of microbiome maturation [4,65,68]. In Bäckhed et al., the cessation of breastfeeding was associated with a shift toward an adult-like composition enriched in *Bacteroides*, *Bilophila*, *Roseburia*, *Clostridium*, and *Anaerostipes* [4]. However, the gut microbiome of infants who were still breastfed at 12 months remained enriched in *Bifidobacterium*, *Lactobacillus*, *Collinsella*, *Megasphaera*, and *Veillonella* [4]. Various models of "microbiota age" showed that 12-month-old infants who were still receiving breast milk appeared younger than those who had been weaned. In another study, the diversity index between infants receiving some breast milk and infants no longer receiving breast milk began to converge over time, likely as a result of reduced proportion of breast milk in the diet [65]. Indeed, infants who are introduced to solid foods before 6 months undergo early maturation of the microbiota, whereas those who continue to receive breast milk after solid food introduction suppress the diversification and enrichment of bacteria typically associated with solid foods [62].

The infant gut microbiota has been shown to change rapidly, within 5 days, during the transition from breastfeeding to cow's milk [69]. The change in diet to solid food is associated with greater bacterial load and increased levels of short-chain fatty acids [70]. Overall, the introduction of solid foods is associated with a rapid shift in the infant microbiota to a mature, adult-like state dominated by species capable of degrading glycans, mucin, and complex carbohydrates as well as producing short-chain fatty acids [71].

Intriguingly, differences in the gut microbiota of individuals who receive breast milk versus those who do not are still evident later in life. In the aforementioned meta-analysis, differences in the gut microbiota between exclusively breastfed infants and those who received other nutrition were still evident between 6 months and 2 years of age [67]. Taken together, these studies imply that early life nutrition plays a key and potentially lasting role in microbiome dynamics.

Hospitalization

In the United States, the vast majority of births occur in hospital settings [72]. Therefore, these unique indoor environments represent a major source of microbial exposure to neonates. It stands to reason that cesarean-delivered infants, as discussed earlier, may be more directly exposed to hospital staff and environmental microbes, but some argue that the place of delivery (home vs hospital) also has an effect on microbiome composition in vaginal delivery [73].

A few studies have examined the microbes present in the hospital environment. For example, operating rooms, the site of delivery for cesarean section infants, are dominated by *Staphylococcus* and *Corynebacterium*, probably from healthcare workers' skin [74]. Furthermore, the microbiome of the operating room dust varied by city and by hospital [74]. Another group characterized the NICU in two hospitals and found a high diversity of human-associated bacteria, including many bacterial taxa known to be opportunistic pathogens and/or skin associated [75].

Microbial differences between hospitals were further supported by a study that compared infants' gut microbiome from two geographically separate NICUs over 2 years [76]. The authors identified differences in microbial succession by hospital, with greater change in microbiota in hospital 1 infants than in hospital 2 infants [76]. Understanding how the microbes in the hospital environment change over time has implications for the seeding of the neonatal gut microbiome.

Some studies have found that interindividual differences between hospitalized infants' microbiota tend to converge over time, associated with length of NICU stay [77]. This coincides with results from a longitudinal study of adult hospitals in which patient skin and room surfaces became more microbially similar over the course of their stay [78]. Various groups have probed the directionality of transfer between the room and the patient, with Lax et al. suggesting that patients initially acquired room-associated taxa that predated their stay but that their own microbial signatures began to influence the room microbiota over time [78]. The idea that hospitals both colonize and are colonized by patients' microbes is further reinforced by imputed source tracking of skin microbes in the NICU or nursery [79].

Brooks et al. identified overlap between bacterial genomes present in NICU rooms and infant guts, showing that strains of *Staphylococcus epidermidis*, *Enterococcus faecalis*, *Pseudomonas aeruginosa*, and *Klebsiella pneumoniae*, which are frequently implicated in nosocomial infection, were present on sinks and surfaces as well as in infant guts [80]. Resolving by time, the authors concluded that nearly all strains associated with infant gut colonization were detectable in the room after, and often before, detection in the gut [80]. The directionality of transfer remains to be fully elucidated, but organisms that appear in the gastrointestinal (GI) tract in the early phase of colonization have reservoirs in the room environment [81]. Therefore, gut microbes could be introduced from room sources, grow in the infant, and be disseminated, resulting in bidirectional colonization [82]. Overall, it seems that hospitals harbor their own microbes and should be considered along with diet, mode of delivery, and antibiotics as a determinant of the early gut microbiome.

Other environmental exposures

In today's industrialized world, we spend most of our time indoors, be it at home, school, or work [83]. The increasing interaction with a confined indoor environment tends to decrease our interaction with outdoor microbial diversity. Microbes and viruses found in soil, dust particles, animals, plants, insects, and various other surfaces also comprise sources of the microbiota in outdoor air [84]. Microbial inhalation through air is a qualitative and quantitative variable which varies depending on an individual's activity pattern and the frequency of exposure, which are determined by factors such as the location (e.g., stable, office, school, outdoor recreational area, residence), and building operation parameters (e.g., ventilation, materials, moisture damage), as well as the diversity of occupants and occupant behavior, and the surrounding outdoor environment (e.g., urban vs rural vs farming). One study found greater bacterial abundance indoors compared to outdoors [85], whereas another study found comparable density both indoors and outdoors [86,87]. The risk of asthma and allergies has been associated with factors that influence microbial exposure, for example, older siblings and farm exposure have been associated with protection from asthma. Older siblings are assumed to provide more microbial exposure due to their school interactions with other children, which increases the likelihood that infants will be exposed to a greater diversity of pathogens, viruses, and other bacterial immune adjuvants; meanwhile, the farming environment represents a cornucopia of diverse microbial exposures likely to stimulate immune activity and potential to alter allergic disease outcomes. In contrast, the presence of pets and visiting day care centers are found with less consistent associations. Remarkably, the different determinants seem to be independent [88,89].

Pets

Sharing a home with a cat or a dog early in life has been associated with reduced development of asthma and allergies [90–92]. Significant protection from asthma with reliable statistical power was observed in a study that included over a million children, indicating that early life home sharing with a dog could protect from asthma in about 1 in 10 children [88]. Another study showed the protective effect of dog interaction on atopic sensitization among children who were generally linked to lower microbial exposures such as those born by cesarean section and those who lack siblings [93]. Several other studies suggest that dog ownership has an influence on the indoor microbiota rather than increased richness and diversity [92,94–100]. However, owning a dog does not influence fungal diversity [99,101], and interestingly cats do not appear to have a significant impact on indoor microbial diversity or allergic sensitization as dogs [94–96,99,100]. The characteristics of a specific dog breed or behavior, such as frequency of contact between indoor and outdoor environments and likelihood of physical interaction with a child, could be speculated to determine the microbial exposures and thereby its influence on asthma and allergies [100,102]. While dogs are associated with an increased diversity of bacteria in the indoor environment, this diversity appears to be associated with their own microbiome rather than due to an increase in bacteria associated with the outdoors [94,97,99]; with Bacteroidetes, Firmicutes, and Fusobacteria, and *Moraxella* and *Porphyromonas* dominating [94,96,97]. Airway colonization by specific *Moraxella* spp. is associated with an increased risk of asthma; however, it also seems that different species of *Moraxella* elicit different host responsiveness [64]. Oral infection of *Porphyromonas gingivalis* in humans has been observed to protect against asthma and airway inflammation [103]. This association leaves an open question if the exposure to such organisms could stem from the dog's microbiome.

Siblings

The hygiene hypothesis posits that early life exposures to certain microorganisms protect children from allergic diseases through training the immune system. One such example is the repeated observation of an inverse association between the number of siblings and hay fever [104–107]. Interestingly, the protective effects were stronger with elder siblings than the younger ones, which indicates that microbial exposure benefits may be limited to the first 2 years of life. Siblings born close in time to each other were found to have stronger sibling effects due to shared microbial exposures [108,109]. An increase in infection rates has been linked with attending day care or having siblings. Therefore, increased microbial exposure appears to modulate immune system responses in infants, reducing the likelihood of allergic disease.

Implications for lifelong health

Perturbations of the microbiome due to these environmental factors may not persist throughout life, but increasingly, evidence suggests that the early life microbiome plays a critical role in patterning immune and metabolic development. For example, a number of groups have examined the influence of the microbiome on the heritability of obesity. Animal studies have demonstrated a microbial etiology for obesity, as transplanting the gut microbial community from obese mice into germ-free mice can lead to increased body fat in the recipient mice [110]. Furthermore, obese adults have different microbial profiles than lean controls, suggesting that particular microbiome compositions can be both obesogenic and transmittable. Given the current understanding of the maternal microbiome as a source for children's colonizers, the vertical transmission of obesogenic microbes represents a possible pathway linking maternal and child weight [111]. Indeed, Tun et al. showed that birth mode and infant gut microbiota mediated the association between maternal prepregnancy weight and childhood weight at ages 1 and 3 years [112]. They demonstrated that infants born vaginally to overweight or obese mothers were three times more likely to become overweight, whereas cesarean-delivered infants of overweight mothers had a fivefold risk of overweight [112]. Galley et al. found higher abundances of *Faecalibacterium* spp., *Eubacterium* spp., *Oscillibacter* spp., and *Blautia* spp. in the stool of toddlers whose mothers were obese compared to those whose mothers were nonobese [111]. Cox et al. treated mice with subtherapeutic antibiotic treatment at weaning and found that, while the microbiota in young mice returned to a normal populations structure after antibiotic cessation, the treatment was associated with elevated adiposity that emerged in early to mid-adulthood [59]. Notably, there were specific organisms consistently underrepresented in the antibiotic treated mice before obesity phenotype development, implying that their loss during a critical window is detrimental [59]. Differences in bacterial abundances might lead to alterations in energy procurement from food. In particular, it has been hypothesized that greater energy harvest from short-chain fatty acids occurs by gut microbes when the abundance of Firmicutes exceeds that of the Bacteroidetes phylum [112]. Taken together, the current literature suggests that perturbations to the early life microbiota have lasting health consequences and suggest an additional mechanism by which maternal metabolic states can influence their children.

Relatedly, the early life microbiota has ramifications for immune system programming as well. One group used a colitis-susceptible mouse model to demonstrate that germ-free adult pregnant mice inoculated with a gut microbial community shaped by antibiotic exposure transmitted the altered microbiota to their offspring with high fidelity [60]. The offspring that received antibiotic perturbed gut microbiota via this vertical transmission developed increased colitis compared with those that received normal gut microbiota [60]. Therefore, we can conclude that maternal antibiotic exposure may shape not only the mother's own gut microbiota, but may also affect her children's long-term health. A recent study in humans compared the metagenomically encoded functions of the microbiome in infants based on their mode of delivery. They linked functions, such as the synthesis of lipopolysaccharide (LPS), to particular strains that were transmitted by vaginal delivery but not by cesarean section, and they found that the LPS was directly stimulatory to primary immune cells and resulted in higher plasma levels of tumor necrosis factor (TNF-α) and interleukin 18 (IL-18) [113]. This supports the idea that cesarean delivery disrupts not only mother-to-neonate transmission of specific microbial strains, but also their associated functions and immunostimulatory potential. Furthermore, Arrieta et al. found that children who developed asthma had reductions in four specific bacterial genera (*Faecalibacterium*, *Lachnospira*, *Veillonella*, and *Rothia*) in the first 100 days of life [48]. Interestingly, these gut microbial changes were less apparent by 1 year of age, suggesting that the window of development is very early in life.

Overall, the literature suggests that even transient perturbations of the microbiome in early life can lead to lasting health consequences later. Much remains to be clarified about the time course and plasticity of this developmental window, and questions persist about how quickly dietary, antibiotic, probiotic, or other microbial interventions can modify the microbiome and what the optimal duration of such therapies might be [47]. However, early life appears to be a time when therapeutic interventions might eventually be deployed to affect long-term health.

Conclusion

The colonization of the human infant by the microbial world and the subsequent successional dynamics of the new microbiome are well evidenced to have a significant impact on childhood disease. The factors that influence colonization and succession toward maturation are manifold and present an extraordinarily complex dynamics. A child's immune system is influenced by his mother's microbiome in utero, and then through microbial exposure during delivery, breastfeeding, and interaction with the world around him. While large cohort studies have started to identify dominant factors, such as vaginal delivery, breastfeeding, pet ownership, and farm exposure, that can explain a large proportion of the variance in childhood allergic and metabolic diseases, we are not yet able to prescribe a course of action to promote a healthy immune system by augmenting microbial exposure. For example, we know that growing up on a farm decreases the risk of asthma development, but this does not mean we can prescribe farm exposure to young children. Questions remain about when a child should be first exposed to a farm, length and frequency of exposure. Clinical interventions require us to provide evidence of a dose-dependent response to a therapy, and while evidence from observational and interventional preclinical and clinical studies is building, we are not yet able to verify dose-dependency for any of the factors identified in this chapter. Future work is aimed currently at dissecting the mechanisms that underpin many of the observed associations. In some cases, it may be possible to identify a small molecule, produced by gut, skin, or oral bacteria, that in some way influences host health. If we can isolate and

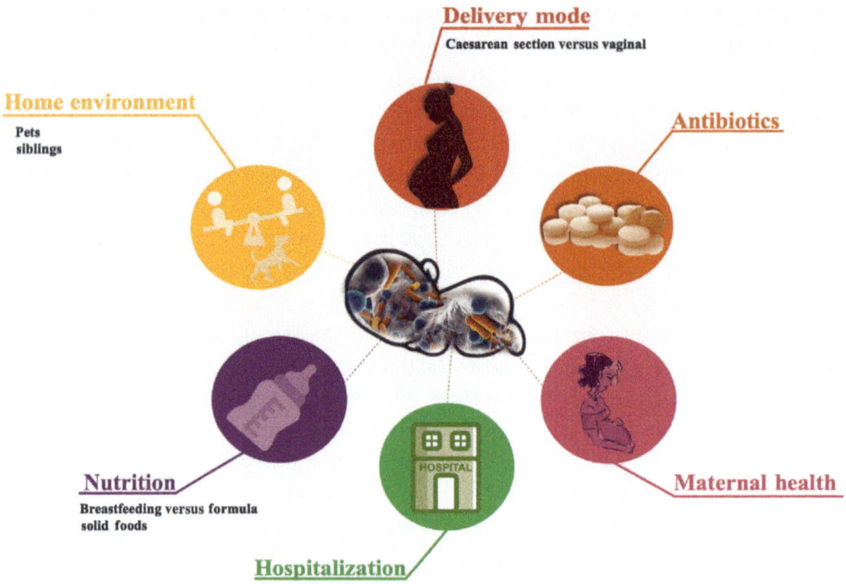

FIGURE 2.1 Multiple factors influence the colonization of the infant microbiome and include the home environment, delivery mode, antibiotics administration at birth or in youth, maternal health, hospitalization, and nutrition.

test that small molecule as a drug in clinical intervention trials, we will be in a strong position to harness the power of microbial exposure through the targeted activities. Meanwhile, increased microbial exposure from birth to 2 years, as long as associated with vaccination and common sense, appears to have benefits that may outweigh risk, but more data is needed to identify targets of therapeutic interventions (Fig. 2.1).

References

[1] Gilbert JA, Lynch SV. Community ecology as a framework for human microbiome research. Nat Med 2019;25:884–9.

[2] Gilbert JA, Quinn RA, Debelius J, Xu ZZ, Morton J, Garg N, et al. Microbiome-wide association studies link dynamic microbial consortia to disease. Nature 2016;535:94–103.

[3] Gilbert JA, Blaser MJ, Caporaso JG, Jansson JK, Lynch SV, Knight R. Current understanding of the human microbiome. Nat Med 2018;24:392–400.

[4] Bäckhed F, Roswall J, Peng Y, Feng Q, Jia H, Kovatcheva-Datchary P, et al. Dynamics and stabilization of the human gut microbiome during the first year of life. Cell Host Microbe 2015;17:852.

[5] Nayfach S, Rodriguez-Mueller B, Garud N, Pollard KS. An integrated metagenomics pipeline for strain profiling reveals novel patterns of bacterial transmission and biogeography. Genome Res 2016;26:1612–25.

[6] Goodrich JK, Davenport ER, Beaumont M, Jackson MA, Knight R, Ober C, et al. Genetic determinants of the gut microbiome in UK twins. Cell Host Microbe 2016;19:731–43.

[7] Rothschild D, Weissbrod O, Barkan E, Kurilshikov A, Korem T, Zeevi D, et al. Environment dominates over host genetics in shaping human gut microbiota. Nature 2018;555:210–15.

[8] Falony G, Joossens M, Vieira-Silva S, Wang J, Darzi Y, Faust K, et al. Population-level analysis of gut microbiome variation. Science 2016;352:560–4.

[9] Mueller NT, Bakacs E, Combellick J, Grigoryan Z, Dominguez-Bello MG. The infant microbiome development: mom matters. Trends Mol Med 2015;21:109–17.

[10] Aloisio I, Mazzola G, Corvaglia LT, Tonti G, Faldella G, Biavati B, et al. Influence of intrapartum antibiotic prophylaxis against group B *Streptococcus* on the early newborn gut composition and evaluation of the anti-*Streptococcus* activity of *Bifidobacterium* strains. Appl Microbiol Biotechnol 2014;98:6051–60.

[11] Keski-Nisula L, Kyynäräinen H-R, Kärkkäinen U, Karhukorpi J, Heinonen S, Pekkanen J. Maternal intrapartum antibiotics and decreased vertical transmission of *Lactobacillus* to neonates during birth. Acta Paediatr 2013;102:480–5.

[12] Milani C, Duranti S, Bottacini F, Casey E, Turroni F, Mahony J, et al. The first microbial colonizers of the human gut: composition, activities, and health implications of the infant gut microbiota. Microbiol Mol Biol Rev 2017;81.

[13] Rodríguez JM, Murphy K, Stanton C, Ross RP, Kober OI, Juge N, et al. The composition of the gut microbiota throughout life, with an emphasis on early life. Microb Ecol Health Dis 2015;26.

[14] Macpherson AJ, de Agüero MG, Ganal-Vonarburg SC. How nutrition and the maternal microbiota shape the neonatal immune system. Nat Rev Immunol 2017;17:508–17.

[15] Collado MC, Isolauri E, Laitinen K, Salminen S. Effect of mother's weight on infant's microbiota acquisition, composition, and activity during early infancy: a prospective follow-up study initiated in early pregnancy. Am J Clin Nutr 2010;92:1023–30.

[16] Ma J, Prince AL, Bader D, Hu M, Ganu R, Baquero K, et al. High-fat maternal diet during pregnancy persistently alters the offspring microbiome in a primate model. Nat Commun 2014;5:3889.

[17] Chu DM, Antony KM, Ma J, Prince AL, Showalter L, Moller M, et al. The early infant gut microbiome varies in association with a maternal high-fat diet. Genome Med 2016;8.

[18] Lundgren SN, Madan JC, Emond JA, Morrison HG, Christensen BC, Karagas MR, et al. Maternal diet during pregnancy is related with the infant stool microbiome in a delivery mode-dependent manner. Microbiome 2018;6:109.

[19] Gosalbes MJ, Llop S, Vallès Y, Moya A, Ballester F, Francino MP. Meconium microbiota types dominated by lactic acid or enteric bacteria are differentially associated with maternal eczema and respiratory problems in infants. Clin Exp Allergy J Br Soc Allergy Clin Immunol 2013;43:198–211.

[20] Hu J, Nomura Y, Bashir A, Fernandez-Hernandez H, Itzkowitz S, Pei Z, et al. Diversified microbiota of meconium is affected by maternal diabetes status. PLoS One 2013;8.

[21] Ferretti P, Pasolli E, Tett A, Asnicar F, Gorfer V, Fedi S, et al. Mother-to-infant microbial transmission from different body sites shapes the developing infant gut microbiome. Cell Host Microbe 2018;24 133 – 145.e5.

[22] Perez-Muñoz ME, Arrieta M-C, Ramer-Tait AE, Walter J. A critical assessment of the "sterile womb" and "in utero colonization" hypotheses: implications for research on the pioneer infant microbiome. Microbiome 2017;5:48.

[23] Rutayisire E, Huang K, Liu Y, Tao F. The mode of delivery affects the diversity and colonization pattern of the gut microbiota during the first year of infants' life: a systematic review. BMC Gastroenterol 2016;16:86.

[24] Salas Garcia MC, Yee AL, Gilbert JA, Dsouza M. Dysbiosis in Children Born by Caesarean Section. Ann Nutr Metab 2018;73:24–32.

[25] Dominguez-Bello MG, Costello EK, Contreras M, Magris M, Hidalgo G, Fierer N, et al. Delivery mode shapes the acquisition and structure of the initial microbiota across multiple body habitats in newborns. Proc Natl Acad Sci 2010;107:11971–5.

[26] Neu J, Rushing J. Cesarean versus vaginal delivery: long term infant outcomes and the hygiene hypothesis. Clin Perinatol 2011;38:321–31.

[27] Azad MB, Konya T, Maughan H, Guttman DS, Field CJ, Chari RS, et al. Gut microbiota of healthy Canadian infants: profiles by mode of delivery and infant diet at 4 months. CMAJ 2013;185.

[28] Jakobsson HE, Abrahamsson TR, Jenmalm MC, Harris K, Quince C, Jernberg C, et al. Decreased gut microbiota diversity, delayed Bacteroidetes colonisation and reduced Th1 responses in infants delivered by caesarean section. Gut 2014;63:559–66.

[29] Del Chierico F, Vernocchi P, Petrucca A, Paci P, Fuentes S, Pratico G, et al. Phylogenetic and metabolic tracking of gut microbiota during perinatal development. PLoS One 2015;10:e0137347.

[30] Hill CJ, Lynch DB, Murphy K, Ulaszewska M, Jeffery IB, O'Shea CA, et al. Evolution of gut microbiota composition from birth to 24 weeks in the INFANTMET Cohort. Microbiome 2017;5.

[31] Yee AL, Miller E, Dishaw LJ, Gordon JM, Ji M, Dutra S, et al. Longitudinal microbiome composition and stability correlate with increased weight and length of very-low-birth-weight infants. mSystems 2019;4.

[32] Dahl C, Stigum H, Valeur J, Iszatt N, Lenters V, Peddada S, et al. Preterm infants have distinct microbiomes not explained by mode of delivery, breastfeeding duration or antibiotic exposure. Int J Epidemiol 2018;.

[33] Grier A, Qiu X, Bandyopadhyay S, Holden-Wiltse J, Kessler HA, Gill AL, et al. Impact of prematurity and nutrition on the developing gut microbiome and preterm infant growth. Microbiome 2017;5:158.

[34] La Rosa PS, Warner BB, Zhou Y, Weinstock GM, Sodergren E, Hall-Moore CM, et al. Patterned progression of bacterial populations in the premature infant gut. Proc Natl Acad Sci U S A 2014;111:12522–7.

[35] Fouhy F, Watkins C, Hill CJ, O'Shea C-A, Nagle B, Dempsey EM, et al. Perinatal factors affect the gut microbiota up to four years after birth. Nat Commun 2019;10:1517.

[36] Chu DM, Ma J, Prince AL, Antony KM, Seferovic MD, Aagaard KM. Maturation of the infant microbiome community structure and function across multiple body sites and in relation to mode of delivery. Nat Med 2017;23:314–26.

[37] Korpela K, Costea P, Coelho LP, Kandels-Lewis S, Willemsen G, Boomsma DI, et al. Selective maternal seeding and environment shape the human gut microbiome. Genome Res 2018;28:561–8.

[38] Nobel YR, Cox LM, Kirigin FF, Bokulich NA, Yamanishi S, Teitler I, et al. Metabolic and metagenomic outcomes from early-life pulsed antibiotic treatment. Nat Commun 2015;6:7486.

[39] Dethlefsen L, Relman DA. Incomplete recovery and individualized responses of the human distal gut microbiota to repeated antibiotic perturbation. Proc Natl Acad Sci U S A 2011;108:4554–61.

[40] Hicks LA, Bartoces MG, Roberts RM, Suda KJ, Hunkler RJ, Taylor TH, et al. US outpatient antibiotic prescribing variation according to geography, patient population, and provider specialty in 2011. Clin Infect Dis Publ Infect Dis Soc Am 2015;60:1308–16.

[41] Lozupone CA, Stombaugh JI, Gordon JI, Jansson JK, Knight R. Diversity, stability and resilience of the human gut microbiota. Nature 2012;489:220–30.

[42] Ng KM, Ferreyra JA, Higginbottom SK, Lynch JB, Kashyap PC, Gopinath S, et al. Microbiota-liberated host sugars facilitate post-antibiotic expansion of enteric pathogens. Nature 2013;502:96–9.

[43] Stefka AT, Feehley T, Tripathi P, Qiu J, McCoy K, Mazmanian SK, et al. Commensal bacteria protect against food allergen sensitization. Proc Natl Acad Sci U S A 2014;111:13145–50.

[44] Hoskin-Parr L, Teyhan A, Blocker A, Henderson AJW. Antibiotic exposure in the first two years of life and development of asthma and other allergic diseases by 7.5 yr: a dose-dependent relationship. Pediatr Allergy Immunol Publ Eur Soc Pediatr Allergy Immunol 2013;24:762–71.

[45] Bailey LC, Forrest CB, Zhang P, Richards TM, Livshits A, DeRusso PA. Association of antibiotics in infancy with early childhood obesity. JAMA Pediatr 2014;168:1063–9.

[46] Kronman MP, Zaoutis TE, Haynes K, Feng R, Coffin SE. Antibiotic exposure and IBD development among children: a population-based cohort study. Pediatrics 2012;130: e794–803.

[47] Tamburini S, Shen N, Wu HC, Clemente JC. The microbiome in early life: implications for health outcomes. Nat Med 2016;22:713–22.

[48] Arrieta M-C, Stiemsma LT, Dimitriu PA, Thorson L, Russell S, Yurist-Doutsch S, et al. Early infancy microbial and metabolic alterations affect risk of childhood asthma. Sci Transl Med 2015;7 307ra152.

[49] Langdon A, Crook N, Dantas G. The effects of antibiotics on the microbiome throughout development and alternative approaches for therapeutic modulation. Genome Med 2016;8:39.

[50] Greenwood C, Morrow AL, Lagomarcino AJ, Altaye M, Taft DH, Yu Z, et al. Early empiric antibiotic use in preterm infants is associated with lower bacterial diversity and higher relative abundance of *Enterobacter*. J Pediatr 2014;165:23–9.

[51] Moore AM, Ahmadi S, Patel S, Gibson MK, Wang B, Ndao MI, et al. Gut resistome development in healthy twin pairs in the first year of life. Microbiome 2015;3:27.

[52] Fouhy F, Guinane CM, Hussey S, Wall R, Ryan CA, Dempsey EM, et al. High-throughput sequencing reveals the incomplete, short-term recovery of infant gut microbiota following parenteral antibiotic treatment with ampicillin and gentamicin. Antimicrob Agents Chemother 2012;56:5811–20.

[53] Bokulich NA, Chung J, Battaglia T, Henderson N, Jay M, Li H, et al. Antibiotics, birth mode, and diet shape microbiome maturation during early life. Sci Transl Med 2016;8:343–82.

[54] Yassour M, Vatanen T, Siljander H, Hämäläinen A-M, Härkönen T, Ryhänen SJ, et al. Natural history of the infant gut microbiome and impact of antibiotic treatment on bacterial strain diversity and stability. Sci Transl Med 2016;8:343–81.

[55] Bender JM, Li F, Purswani H, Capretz T, Cerini C, Zabih S, et al. Early exposure to antibiotics in the neonatal intensive care unit alters the taxonomic and functional infant gut microbiome. J Matern-Fetal Neonatal Med J Eur Assoc Perinat Med Fed Asia Ocean Perinat Soc Int Soc Perinat Obstet 2019;1–9.

[56] Korpela K, Salonen A, Virta LJ, Kekkonen RA, Forslund K, Bork P, et al. Intestinal microbiome is related to lifetime antibiotic use in Finnish pre-school children. Nat Commun 2016;7.

[57] Zaura E, Brandt BW, Teixeira de Mattos MJ, Buijs MJ, Caspers MPM, Rashid M-U, et al. Same exposure but two radically different responses to antibiotics: resilience of the salivary microbiome versus long-term microbial shifts in feces. mBio 2015;6 e01693 − 01615.

[58] Ruiz VE, Battaglia T, Kurtz ZD, Bijnens L, Ou A, Engstrand I, et al. A single early-in-life macrolide course has lasting effects on murine microbial network topology and immunity. Nat Commun 2017;8.

[59] Cox LM, Yamanishi S, Sohn J, Alekseyenko AV, Leung JM, Cho I, et al. Altering the intestinal microbiota during a critical developmental window has lasting metabolic consequences. Cell 2014;158:705–21.

[60] Schulfer A, Blaser MJ. Risks of antibiotic exposures early in life on the developing microbiome. PLoS Pathog 2015;11.

[61] Fouhy F, Ross RP, Fitzgerald GF, Stanton C, Cotter PD. Composition of the early intestinal microbiota: knowledge, knowledge gaps and the use of high-throughput sequencing to address these gaps. Gut Microbes 2012;3:203–20.

[62] Pannaraj PS, Li F, Cerini C, Bender JM, Yang S, Rollie A, et al. Association between breast milk bacterial communities and establishment and development of the infant gut microbiome. JAMA Pediatr 2017;171:647–54.

[63] Jost T, Lacroix C, Braegger CP, Rochat F, Chassard C. Vertical mother-neonate transfer of maternal gut bacteria via breastfeeding. Env Microbiol 2014;16:2891–904.

[64] Yatsunenko T, Rey FE, Manary MJ, Trehan I, Dominguez-Bello MG, Contreras M, et al. Human gut microbiome viewed across age and geography. Nature 2012;486.

[65] Stewart CJ, Ajami NJ, O'Brien JL, Hutchinson DS, Smith DP, Wong MC, et al. Temporal development of the gut microbiome in early childhood from the TEDDY study. Nature 2018;562:583–8.

[66] Vatanen T, Franzosa EA, Schwager R, Tripathi S, Arthur TD, Vehik K, et al. The human gut microbiome in early-onset type 1 diabetes from the TEDDY study. Nature 2018;562:589–94.

[67] Ho NT, Li F, Lee-Sarwar KA, Tun HM, Brown BP, Pannaraj PS, et al. Meta-analysis of effects of exclusive breastfeeding on infant gut microbiota across populations. Nat Commun 2018;9:4169.

[68] Bergstrom A, Skov TH, Bahl MI, Roager HM, Christensen LB, Ejlerskov KT, et al. Establishment of intestinal microbiota during early life: a longitudinal, explorative study of a large cohort of Danish infants. Appl Env Microbiol 2014;80:2889–900.

[69] Davis MY, Zhang H, Brannan LE, Carman RJ, Boone JH. Rapid change of fecal microbiome and disappearance of *Clostridium difficile* in a colonized infant after transition from breast milk to cow milk. Microbiome 2016;4.

[70] Koenig JE, Spor A, Scalfone N, Fricker AD, Stombaugh J, Knight R, et al. Succession of microbial consortia in the developing infant gut microbiome. Proc Natl Acad Sci U S A 2011;108(Suppl 1):4578–85.

[71] Robertson RC, Manges AR, Finlay BB, Prendergast AJ. The human microbiome and child growth – first 1000 days and beyond. Trends Microbiol 2018;.

[72] MacDorman MF, Declercq E. Trends and state variations in out-of-hospital births in the United States, 2004–2017. Birth Berkeley Calif 2019;46:279–88.

[73] van Nimwegen FA, Penders J, Stobberingh EE, Postma DS, Koppelman GH, Kerkhof M, et al. Mode and place of delivery, gastrointestinal microbiota, and their influence on asthma and atopy. J Allergy Clin Immunol 2011;128:948–55 e1–3.

[74] Shin H, Pei Z, Martinez KA, Rivera-Vinas JI, Mendez K, Cavallin H, et al. The first microbial environment of infants born by C-section: the operating room microbes. Microbiome 2015;3:59.

[75] Hewitt KM, Mannino FL, Gonzalez A, Chase JH, Caporaso JG, Knight R, et al. Bacterial diversity in two neonatal intensive care units (NICUs). PLoS One 2013;8:e54703.

[76] Taft DH, Ambalavanan N, Schibler KR, Yu Z, Newburg DS, Ward DV, et al. Intestinal microbiota of preterm infants differ over time and between hospitals. Microbiome 2014;2:36.

[77] Patel AL, Mutlu EA, Sun Y, Koenig L, Green S, Jakubowicz A, et al. Longitudinal survey of microbiota in hospitalized preterm very-low-birth-weight infants. J Pediatr Gastroenterol Nutr 2016;62:292–303.

[78] Lax S, Sangwan N, Smith D, Larsen P, Handley KM, Richardson M, et al. Bacterial colonization and succession in a newly opened hospital. Sci Transl Med 2017;9.

[79] Younge NE, Araújo-Pérez F, Brandon D, Seed PC. Early-life skin microbiota in hospitalized preterm and full-term infants. Microbiome 2018;6:98.

[80] Brooks B, Olm MR, Firek BA, Baker R, Thomas BC, Morowitz MJ, et al. Strain-resolved analysis of hospital rooms and infants reveals overlap between the human and room microbiome. Nat Commun 2017;8:1–7.

[81] Brooks B, Firek BA, Miller CS, Sharon I, Thomas BC, Baker R, et al. Microbes in the neonatal intensive care unit resemble those found in the gut of premature infants. Microbiome 2014;2:1.

[82] Brooks B, Olm MR, Firek BA, Baker R, Geller-McGrath D, Reimer SR, et al. The developing premature infant gut microbiome is a major factor shaping the microbiome of neonatal intensive care unit rooms. Microbiome 2018;6:112.

[83] Klepeis NE, Nelson WC, Ott WR, Robinson JP, Tsang AM, Switzer P, et al. The National Human Activity Pattern Survey (NHAPS): a resource for assessing exposure to environmental pollutants. J Expo Sci Env Epidemiol 2001;11:231.

[84] Miletto M, Lindow SE. Relative and contextual contribution of different sources to the composition and abundance of indoor air bacteria in residences. Microbiome 2015;3:61.

[85] Hospodsky D, Yamamoto N, Nazaroff WW, Miller D, Gorthala S, Peccia J. Characterizing airborne fungal and bacterial concentrations and emission rates in six occupied children's classrooms. Indoor Air 2015;25:641–52.

[86] Prussin II AJ, Garcia EB, Marr LC. Total virus and bacteria concentrations in indoor and outdoor air. Env Sci Technol Lett 2015;2:84.

[87] Kembel SW, Jones E, Kline J, Northcutt D, Stenson J, Womack AM, et al. Architectural design influences the diversity and structure of the built environment microbiome. ISME J 2012;6:1469.

[88] Fall T, Lundholm C, Örtqvist AK, Fall K, Fang F, Hedhammar Å, et al. Early exposure to dogs and farm animals and the risk of childhood asthma. JAMA Pediatr 2015;169 e153219–e153219.

[89] Genuneit J, Strachan DP, Büchele G, Weber J, Loss G, Sozanska B, et al. The combined effects of family size and farm exposure on childhood hay fever and atopy. Pediatr Allergy Immunol 2013;24:293–8.

[90] Carlsen KCL, Roll S, Carlsen K-H, Mowinckel P, Wijga AH, Brunekreef B, et al. Does pet ownership in infancy lead to asthma or allergy at school age? Pooled analysis of individual participant data from 11 European birth cohorts. PLoS One 2012;7:e43214.

[91] Lodge CJ, Allen KJ, Lowe AJ, Hill DJ, Hosking CS, Abramson MJ, et al. Perinatal cat and dog exposure and the risk of asthma and allergy in the urban environment: a systematic review of longitudinal studies. Clin Dev Immunol 2011;2012.

[92] Sitarik AR, Havstad S, Levin AM, Lynch SV, Fujimura KE, Ownby DR, et al. Dog introduction alters the home dust microbiota. Indoor Air 2018;28:539–47.

[93] Wegienka G, Havstad S, Kim H, Zoratti E, Ownby D, Woodcroft KJ, et al. Subgroup differences in the associations between dog exposure during the first year of life and early life allergic outcomes. Clin Exp Allergy 2017;47:97–105.

[94] Dunn RR, Fierer N, Henley JB, Leff JW, Menninger HL. Home life: factors structuring the bacterial diversity found within and between homes. PLoS One 2013;8: e64133.

[95] Ownby DR, Peterson EL, Wegienka G, Woodcroft KJ, Nicholas C, Zoratti E, et al. Are cats and dogs the major source of endotoxin in homes? Indoor Air 2013;23:219–26.

[96] Lee MK, Carnes MU, Butz N, Azcarate-Peril MA, Richards M, Umbach DM, et al. Exposures related to house dust microbiota in a US farming population. Env Health Perspect 2018;126:67001.

[97] Kettleson EM, Adhikari A, Vesper S, Coombs K, Indugula R, Reponen T. Key determinants of the fungal and bacterial microbiomes in homes. Env Res 2015;138:130–5.

[98] Heinrich J, Gehring U, Douwes J, Koch A, Fahlbusch B, Bischof W, et al. Pets and vermin are associated with high endotoxin levels in house dust. Clin Exp Allergy 2001;31:1839–45.

[99] Parajuli A, Grönroos M, Siter N, Puhakka R, Vari HK, Roslund MI, et al. Urbanization reduces transfer of diverse environmental microbiota indoors. Front Microbiol 2018;9:84.

[100] Fujimura KE, Johnson CC, Ownby DR, Cox MJ, Brodie EL, Havstad SL, et al. Man's best friend? The effect of pet ownership on house dust microbial communities. J Allergy Clin Immunol 2010;126:410.

[101] Nygaard AB, Charnock C. Longitudinal development of the dust microbiome in a newly opened Norwegian kindergarten. Microbiome 2018;6:159.

[102] Barberán A, Dunn RR, Reich BJ, Pacifici K, Laber EB, Menninger HL, et al. The ecology of microscopic life in household dust. Proc R Soc B Biol Sci 2015;282:20151139.

[103] Card JW, Carey MA, Voltz JW, Bradbury JA, Ferguson CD, Cohen EA, et al. Modulation of allergic airway inflammation by the oral pathogen Porphyromonas gingivalis. Infect Immun 2010;78:2488–96.

[104] Strachan DP. Hay fever, hygiene, and household size. BMJ 1989;299:1259.

[105] Strachan DP, Aït-Khaled N, Foliaki S, Mallol J, Odhiambo J, Pearce N, et al. Siblings, asthma, rhinoconjunctivitis and eczema: a worldwide perspective from the International Study of Asthma and Allergies in Childhood. Clin Exp Allergy 2015;45:126–36.

[106] Karmaus W, Botezan C. Does a higher number of siblings protect against the development of allergy and asthma? A review. J Epidemiol Community Health 2002;56:209–17.

[107] Goldberg S, Israeli E, Schwartz S, Shochat T, Izbicki G, Toker-Maimon O, et al. Asthma prevalence, family size, and birth order. Chest 2007;131:1747–52.

[108] Ponsonby A-L, Couper D, Dwyer T, Carmichael A. Cross sectional study of the relation between sibling number and asthma, hay fever, and eczema. Arch Dis Child 1998;79:328–33.

[109] Strachan DP, Taylor EM, Carpenter RG. Family structure, neonatal infection, and hay fever in adolescence. Arch Dis Child 1996;74:422–6.

[110] Turnbaugh PJ, Ley RE, Mahowald MA, Magrini V, Mardis ER, Gordon JI. An obesity-associated gut microbiome with increased capacity for energy harvest. Nature 2006;444:1027–31.

[111] Galley JD, Bailey M, Kamp Dush C, Schoppe-Sullivan S, Christian LM. Maternal obesity is associated with alterations in the gut microbiome in toddlers. PLoS One 2014;9.
[112] Tun HM, Bridgman SL, Chari R, Field CJ, Guttman DS, Becker AB, et al. Roles of birth mode and infant gut microbiota in intergenerational transmission of overweight and obesity from mother to offspring. JAMA Pediatr 2018;172:368–77.
[113] Wampach L, Heintz-Buschart A, Fritz JV, Ramiro-Garcia J, Habier J, Herold M, et al. Birth mode is associated with earliest strain-conferred gut microbiome functions and immunostimulatory potential. Nat Commun 2018;9:1–14.

Chapter 3

The microbiome and immune system development

Modupe O. Coker[1,2] and Juliette C. Madan[2,3]
[1]*Department of Oral Biology, Rutgers School of Dental Medicine, Newark, NJ, United States,* [2]*Department of Epidemiology, Geisel School of Medicine at Dartmouth, Hanover, NH, United States,* [3]*Department of Pediatrics, Dartmouth Hitchcock Medical Center, Lebanon, NH, United States*

Innate and adaptive immunity defined

At the time of birth, neonates rely on *innate immunity*, or the cellular and extracellular components with which they are born, and do not rely on pathogen presentation, exposures, or immune learning, in order to protect themselves against pathogens [1–3]. Cellular components of innate immunity include dendritic cells, neutrophils, macrophages, monocytes, and natural killer (NK) cells; extracellular components of innate immunity include complement, fibronectin, antimicrobial peptides and proteins, and acute phase proteins [4]. The *adaptive immune system* is immature in the first weeks and months of life and is composed of cellular (T cell-mediated) immunity and humoral immunity (B cells and immunoglobulins). In newborns, cellular immunity is dominated by T helper type 2 (Th2) responses, and together with T regulatory cells, secrete cytokines in response to pathogens. Passive immunoglobulins from maternal circulation are acquired during fetal life in the third trimester, and as newborn humoral immunity begins to mature, passive immunity from maternal circulation wanes, usually in the first 3 months of postnatal life [2,5–7].

Introduction

There is overwhelming evidence that the infant gut microbiome plays a fundamental role in immune development in the perinatal and neonatal period, assisting in directing the development of the intestines and in both innate and adaptive immune training [8,9]. The intrauterine environment provides initial stimulation and signaling for the fetal innate immune system, and

The Developing Microbiome. DOI: https://doi.org/10.1016/B978-0-12-820602-7.00003-9

passive immunity is provided from maternal circulation particularly in the third trimester. The maternal intestinal microbiome and its metabolic signatures impact fetal immune development and ultimately transfer many of its communities through the process of birth [10,11]. At birth, the infant is a relative *tabula rasa*, or blank slate, for both the immune system and the microbiome when compared with its ultimate maturity in the years following birth. In the first weeks and months of life, the innate and adaptive immune system and the infant's microbiome will coevolve, initiating a bidirectional interplay ultimately resulting in competence and maturity [12,13].

After birth the intestinal epithelial cells mature, fortify tight junctions, and the mucin barrier is formed. The gut-associated lymphoid tissue (GALT) and the multitudes of immune cells evolve, ultimately harboring the majority of the immune cells in the body. The local and systemic functions of the immune cells are, in great part, a result of stimulation by the gut microbiome, which includes bacteria, fungi, viruses, and other eukaryotic species. The introduction to microbes via environmental exposures, such as through vertical transmission from delivery mode or diet, occurs before and during the course of immune maturation [14]. The coevolution of the microbiota and immune system in infant development is the hallmark of homeostatic balance between active response to pathogens and tolerance to harmless and beneficial antigens. Infant immune function differs from older children and adults in that the innate immune system (characterized by low cytokine production) is typically blunted at birth in its immaturity, allowing tolerance to the invasion of trillions of microbes [12,15,16]. The cellular components of the adaptive immune system, including neutrophils, monocytes, macrophages, and dendritic cells, increased in number during fetal life; however, their functionality is impaired in newborns [17], with immature innate signaling and diminished cytokine responses [18].

Gut microbial communities sensitize the immune responses as they build microbial networks with signature functional metabolic pathways. The interdependency between microbes and the immune system is elegantly influenced by maternal breast milk in its provision of passive immune components, prebiotics, and milk microbes that provide substrate for the production of metabolites such as short-chain fatty acids (SCFAs), playing pivotal roles in enhancing the generation of regulatory T cells [19]. Furthermore, the time period during which the microbiome is being established, typically in first 1–2 years of life, coincides with the critical window during which children receive most of their immunizations. This vaccine–immune training relationship models how the colonization of gut microbes can drive infants' capacities to mount beneficial immune responses. Intriguing sex differences in immunity and autoimmune diseases are beginning to be elucidated, and some animal models have clarified that these differences are due in part to sex-specific differences in the intestinal microbiome; findings that warrant further exploration as interventions and immune-mediated prevention strategies are explored [20,21].

In this chapter, the current insights into the role of the microbiome in immune ontogeny are reviewed, highlighting how the development of innate and adaptive immunity in infancy is driven by microbe colonization via initial exposures, particularly maternal-to-infant transfer via delivery mode, and through the provision of breast milk (both the milk microbiota and a multitude of bioactive factors). Animal models that have demonstrated mechanisms behind microbe-driven immune maturation are reviewed, demonstrating how bacterial composition, microbial networks, function, and metabolite production influence critical functions of the immune system in health and in immune-mediated disease. The impact of microbe acquisition on the development of host defense and metabolic function will be summarized to underscore its role in the expression of immune and metabolically mediated disease during childhood. In addition, the likelihood of a critical window for microbe–immune interaction is underscored with exploration of disease outcomes in early life related to aberrant immune responses in order to develop a clear understanding of the cross talk between the microbiome and the burgeoning immune system in order to leverage novel interventions to optimize health for a lifetime.

The immature immune system in infancy

At birth, the neonatal microbiome and the neonate's immune system are underdeveloped, and the neonate is particularly vulnerable to infection [6]. In the first weeks and months of life, the innate and adaptive immune system and the infant's microbiome will coevolve, initiating a bidirectional cross talk (Fig. 3.1) [12,13]. The functional activities of the mucosal immune system are rapidly stimulated at birth by bacterial colonization of the intestines [22]. The intestinal tract prevents invasion by pathogens while tolerating colonization with trillions of commensals required for the metabolism of food, production of vitamins, and production of immune-mediating molecules such as SCFAs [23]. The initial stance of the neonatal immune system is one of tolerance, stemming from fetal training and the preparation for exposures following delivery. The innate immune cells (neutrophils, monocytes) in infancy are represented in lower numbers and with contained activity compared with those of older children and adults, whereas innate lymphoid cells (ILCs) are more active in neonates [24]. The immune system is evolving in a step wise manner over time, beginning in fetal life, and for the neonate, the innate immune system is the first line of defense [18]. In utero, T helper type 2 (Th2) cell responses (demonstrated by a decrease in major histocompatibility complex (MHC) class II molecules—a group of genes responsible for antigen presentation—and diminished inflammatory responses) are promoted, and this persists into early infancy, making infants at higher risk for infections [6]. Neutrophils, macrophages, and dendritic cells (monocytes) function less robustly, and the complement system is incomplete in its function in the neonate [25]. NK cells (lymphoid cells that provide innate

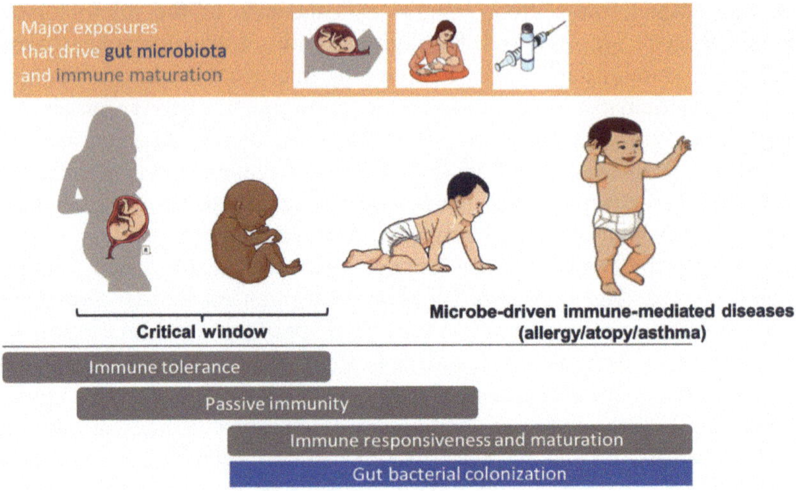

FIGURE 3.1 The infant gut microbiome plays a fundamental role in innate and adaptive immune development in the perinatal and neonatal period. In the first months of life, the immune system and the infant's microbiome will coevolve, initiating a bidirectional interplay ultimately resulting in competence and maturity. There are major events occurring in immune system and gut development at this time. Exposures to maternal vertical transfer of microbes and breast milk drive colonization patterns and the developing immune system. Mechanisms behind microbiome and immune maturation are demonstrated in animal and vaccine models, and supported in longitudinal human studies of microbiome-associated immune-mediated health outcomes. Clarifying the cross talk between the microbiome and the immune system allows opportunities to understand the interactions between host and pathogen in order to leverage novel interventions to optimize health.

immunity against tumors and viruses) are suppressed in the pre-and perinatal period to protect the fetus during development [26]. ILCs, first described in 2010, mediate pro- and antiinflammatory responses, intersecting both the innate and adaptive immune system, and are present in human milk [24].

The GALT is one of the largest immune organs in the body and consists of organized lymphoid tissues, such as mesenteric lymph nodes and Peyer's patches, and more diffusely scattered lymphocytes in the intestinal epithelium; together these tissues substantially impact local and systemic immunity and inflammation [20]. The physical cellular barrier between the gut lumen and the infant's blood stream includes a single layer of epithelial cells and their tight junctions covered by a mucus layer, Paneth and goblet cells, lamina propria, and Peyer's patches with their dendritic cells, macrophages, and neutrophils [16,27]. Beneath the intestinal epithelium lies the largest proportion of immune cells in the human body; nearly 80% of all immune cells in adults are housed in the intestines [23]. Specific sections of intestinal epithelia provide areas for specialized cells to sample luminal content and transport metabolites to Peyer's patches for recognition by the immune system, and dendritic cells and macrophages sample antigens in the gut lumen. Dendritic cells migrate to Peyer's patches or to mesenteric lymph nodes to interact with lymphocytes [28], whereas macrophages stay in the lamina propria communicating locally in the intestines [29]. Cytokines in the intestine are induced by antigens and regulate intestinal immunity.

The ontogeny of innate and adaptive immunity is predicated on protecting the host from unfamiliar or harmful cells, molecules, and organisms throughout the life span while exhibiting self-tolerance. In general, the developing fetus is principally taught immune tolerance of commensal organisms throughout gestation [30,31]. In utero, the immune system is built by alloantigen recognition between the maternal and fetal immune systems and by exposure to immune stimulatory molecules including food antigens, amniotic fluid, maternal cells, and microbes during gestation [6,32–34]. Maternal antibodies (primarily immunoglobulin G) begin to cross the placenta in the first trimester, however, most antibodies are transferred in the last weeks of pregnancy, evidenced by lower serum IgG in preterm infants [35]. Worthy of note is the growing body of evidence highlighting the mechanisms by which alterations in maternal intestinal microbiota and its associated metabolites stimulate and regulate fetal immune development and ultimately the neonatal intestinal microbiota colonization patterns via vertical transmission of microbes [10,11,36–38]. De Aguero et al. provided an example of potential mechanisms of the maternal microbiota stimulating fetal immune training in a murine model, whereby transient intestinal colonization of a pregnant mouse with a mutant strain of *Escherichia coli* was observed to exert effects on the immune system by increasing innate immune cells (impacting ILCs and F4/80(+)CD11c(+) mononuclear cells) in offspring [10]. Another study observed that when pregnant mice were administered vancomycin, infant immunity was altered, as demonstrated by significantly higher lymphocyte numbers compared to controls. This increase in cellularity was attributed to elevated numbers of both CD4+ T cells and B cells, most notably follicular B cells [11].

Early life exposures shape infant immunity and are mediated through the microbiome

At birth, with sudden exposure to maternal and environmental microbes in a delivery-dependent manner, followed by dietary exposures, innate and adaptive immune responses appropriate for early life are activated. The first major microbial exposure is during passage through the birth canal or skin for operative deliveries, and then the subsequent contact with the breast, skin, and the environment. Delivery mode significantly impacts the baseline composition of the infant gut microbial community with higher levels of *Clostridium perfringens* and lower levels of *Bacteroides fragilis* and *Bifidobacterium* in Cesarean-born infants compared to vaginally born infants [39–41].

Infant feeding after birth (breast milk or formula) represents the next major factor that shapes the infant gut microbiome, and human milk is an important source of nutrition and bioactive factors with multifunctional components that shape the developing immune system [42,43]. The study of the developing intestinal microbiome and, in parallel, the immune system would be incomplete without describing the unique and seminal impact of human milk on immune development, both directly and indirectly through its impact on the developing infant intestinal microbiome. Breast milk, particularly colostrum, containing live microbes and metabolites, IgA, immune cells, and cytokines, delivers protective assistance in the form of antibodies, macrophages, neutrophils, lymphocytes, stem cells, and other immune-related factors [44]; more mature milk contains human milk stem cells and antimicrobial properties such as sIgA, TLRs, B-defensin-1, lactoferrin, and CF14 cells for nonspecific immune protection [45]. Infants ingest billions of maternal cells, dominated by macrophages thought to originate, interestingly, from the maternal intestinal immune system [46]. Maternal milk leukocytes (myeloid and lymphoid), which originate from maternal intestinal Peyer's patches, shape the newborn mucosal lymphoid tissues [47]. Overall, these exposures play significant roles in the maturation of an existing, but underdeveloped, immune system.

Breast milk also provides chemical support to the infant's immune system through oligosaccharides, lactoferrin, mucins, and lactadherins, which bind pathogens and antiinflammatory components such as cytokines and growth factors [45]. The breast milk microbiota shapes the developing infant intestinal microbiome, and human milk oligosaccharides (HMOs) act as prebiotics supporting and facilitating the growth of beneficial microbes in the gut by providing substrate for their growth as they populate the gut [42]. In addition, oligosaccharides play a significant role in the developing gut microbiota by directly binding to pathogens in their role in the chemical barrier of the gut and affect molecular pattern signaling pathways such as TLR3, TLR5, IL-1B, and decrease cytokine production and inflammatory responses [48,49]. Human milk, once thought to be sterile, actually contains billions of microbes and is dominated by facultative anaerobes including *Staphylococcus* and *Streptococcus* as well as *Lactobacillus*, *Bifidobacterium*, *Enterococcus*, *Rothia*, and *Propionibacterium* [50–52]. There is significant intraindividual variability in the milk microbiome over the course of lactation [53], and significant interindividual variability in the milk microbiome has shown to be associated with maternal BMI, gestational weight gain, maternal diet, and geographic area in several studies [51,52,54]. The milk microbiota directly shapes the infant intestinal microbiota as microbes in the intestines utilize glycosaminoglycans, glycoproteins, lipids, and oligosaccharides as prebiotics to facilitate growth [55,56]. The early colonizers, including *Bacteroides* and *Bifidobacterium*, engage in a bidirectional interplay with the host, inducing glycan production on the mucosa, which supports microbial growth, and the microbes convert indigestible milk glycans to absorbable SCFAs, which have many immunomodulatory effects [55]. All of these factors act synergistically to shape the breastfed infant microbiota and the infant host response to these microbes.

As an example, HMOs in mother's milk promote the expansion of specific components of the microbiota such as *Bifidobacterium* [57,58], thus contributing to neonatal immune imprinting by influencing the nature of the immune response toward such commensal antigens [38]. Research in infant monkey models demonstrates that breast milk results in increased stability of the developing microbiota and stronger T helper 17 cells development in the first year of life compared with formula-fed animals [59].

Innate and adaptive immunity requires coevolution with the developing microbiome

The innate immune system is typically blunted at birth (evidenced by low production of inflammatory cytokines) to promote the tolerance of the rapidly changing environment and invasion of commensal bacteria. This environment ensures that establishment of the developing microbiota occurs without overt inflammation. This, however, makes the newborn, and particularly the premature baby, susceptible to infections. Cells involved in the innate immune system—neutrophils, monocytes, macrophages, and dendritic cells—which all interact with the adaptive (specific) immune system, increase in abundance throughout gestation [60]. However, the functions of these cells are impaired and inadequate at birth when compared with older children and adults [17]. Specifically, the presence of immature monocytes and macrophages [18] results in impaired innate signaling pathways and diminished cytokine responses. The rise in secretion of antimicrobial peptides [61] by neutrophils and epithelial cells during the neonatal period has been attributed to gene induction by components or the gut microbiome [60]. Several studies have demonstrated impaired cytokine responses, including enhanced production of IL-6, a cytokine with antiinflammatory and Th2-polarizing properties [62,63]. The newborn responds to pathogens and infectious risks with its innate immune responses, primarily the 10 toll-like receptors (TLRs), 1–10, that provide rapid responses to threats. These TLRs act as a main pattern recognition receptors to several pathogen-associated molecular patterns (PAMPs) of microbes on entry into the host [64].

Similarly, the adaptive immunity is not yet developed in infancy, and unlike the innate immune system, it requires education of T and B cells by antigen exposures and formation of memory cells. Cell-mediated immunity is conferred predominantly by T cells, as lymphocytes with specific receptors expand and produce targeted antibodies and cytotoxic T cells [24]; in neonates this process is immature. Neonates have a Th2-dominated cell responses when their immune system is in its immature state; maturation of the Th2 dominance occurs via interaction with bacteria during the early neonatal period, driving barrier functions and Treg and T helper cell balance during a short window of time in early life [65,66]. However, T cells in infancy differ significantly from those in adults in that the fetal immune system, with its Th2 bias, results in suppression of inflammation and elevated levels of IL-10 and TGFβ. This predilection to Th2 is likely in place to avoid pro-inflammatory Th1-type alloimmune responses to maternal tissues that may trigger self-destructive activities [67]. B cells are typically present in secondary lymphoid organs and in the bone marrow, where they contribute to the adaptive immune system via humoral immunity involving the production of antibodies. Antibodies react against invaders by adhering to bacteria, leading to a cascade of events preventing their growth and survival. From birth to 12 weeks of age, maturation of B cells reach their peak [68]. IgM-containing cells are the most prominent up to the first month of life, thereafter IgA-containing cells increase until 2 years of age [69,70]. Most antibody responses are dependent on T-cell interaction as they rely on the engagement of receptors on T helper cells with their corresponding binding partners on antigen-specific B cells. However, B cells demonstrate low levels of these receptors, limiting their capacity to respond [18].

Mechanistic studies have clarified the microbiome's seminal role in immune competence

Germ-free murine models have clarified that competent immune development requires interaction with the developing gut microbes in early life [71,72]. The interaction between intestinal microbes and the developing immune system is likely bidirectional, with the complex intestinal immune system directing colonization [73]. The intestinal microbiota has been shown to influence immune cell activity in several animal models [74], specifically, by comparing antibiotic-treated germ-free to wild-type mice, where microbiota-driven changes in the immune response and immune cell populations were observed [75–78]. Studies of germ-free mice [79,80] indicate that, in the absence of a microbiome, mammals have compromised immune function with impaired development of the GALT, with smaller Peyer's patches and fewer plasma and intraepithelial cells [71,81,82] and impaired secretion of antibodies (including IgA) and antimicrobial peptides [14,83]. These deficiencies are corrected by recolonization with a health-associated mouse commensal microbiota. *Bifidobacterium infantis* has been shown to restore Th1 responses in neonatal mice [84], and serum levels of IgA increase in piglets colonized postpartum [85]. Colonization with *Lactobacillus* has been shown to increase the numbers of IgA-containing B cells [86]. Furthermore, findings from an experiment of inoculated germ-free mice [87] suggest that gut microbiota stimulate the Treg cell development in an IL-2 dependent manner. Certain Clostridia species are associated with increased numbers of Treg cells in murine models, whereas segmented filamentous bacteria are associated with development of the T helper 17 cell lineage in the murine small intestine [72,88].

Human studies, although fewer in number, support the mechanistic findings from animal studies. Data suggest that early colonization with specific bacteria influences systemic immune functions. Specifically, colonization with *B. fragilis* appears to increase the number of circulating IgA and IgM antibody-producing cells in healthy infants [89], and probiotic administration at birth was associated with higher plasma levels of C-reactive protein, total IgA, total IgE, and IL-10 at 6 months compared to controls [90]. Conversely, dysregulation of immune functions in immunodeficient states such as HIV has been linked to microbiome dysbiosis [91].

The metabolome and immune training

Small-molecule products (metabolites) collectively called the "metabolome" mirror gut microbial composition and play a significant role in modifying immune functions. In addition to PAMPs (described earlier) such as bacterial flagellin (recognized by TLR5), lipoteichoic acid from gram-positive bacteria (recognized by TLR2), and peptidoglycan (recognized by TLR2), there is emerging evidence that the immune system can sense various microbial products. Recognition of these small molecules can influence the host immune response [72]. In particular, SCFAs, primarily acetate, propionate, and butyrate, are metabolites produced through fermentation of fibers by microbes. SCFAs have been shown to induce differentiation of Tregs in the colon in various mouse models [92]. Furthermore, butyrate modulates the function of intestinal macrophages leading to a downregulation of lipopolysaccharide-induced pro-inflammatory mediators, including nitric oxide, IL-6, and IL-12, independent of TLR signaling [93]. As another example, a sequence variant (SV) of *Anaerostipes hadrus* encodes the biosynthesis of butyrate shown to decrease the risk of metabolic disease in the host, potentially explaining the difference in body weight between individuals carrying such microbial SVs and those without [94]. A polysaccharide from *B. fragilis*, called *polysaccharide A* [95] has been shown to restore the proportion of splenic CD4-T cells [96,97]. This is further supported by data from Ramakrishna et al. that suggest that PSA administration increases IL-10 production [98].

Microbe–immune interactions occur in a critical window of time in infancy impacting lifelong health

Epidemiological studies in humans and their associated mechanistic studies in animal models have clarified that the interaction of the developing microbiota with the developing immune system occurs in a critical window of time in early infancy. Studies have identified that disturbance of this parallel development results in immune-mediated diseases, such as allergies, atopy, and asthma with lifelong consequences in diseases such as type 1 diabetes [6,99–110]. Elegant murine models have identified that the reverse is true: if perturbations in the parallel development of the microbiome and immunity are corrected within this critical window of time, disease outcomes can be ameliorated [99,111]. Neonatal murine models have identified that the microbial antigens influence preimmune B cell populations (primarily immature B cells that have not undergone somatic mutation) contributing to the development of the B cell repertoire until the age of weaning [112]. Rabbit and chicken models have clarified a specific window of time in neonatal life during which microbial colonization of the gut contributes to specificity of early B cell populations in the GALT, and in murine models B cell receptor editing occurs in the lamina propria at the time of weaning [113]. In this window between delivery and weaning, maternal antibody titers and microbiome-driven B cell maturation occur, providing a window of time during which the gut microbiome may be manipulated to optimize health or improve vaccine efficacy [18,114–117].

Infant microbiome is associated with immune-mediated outcomes and disease in infancy

Description of microbiome-associated immune-mediated disease outcomes would be incomplete without consideration of the "hygiene or microbiota hypothesis" that theorizes that the rise in allergic diseases is a consequence of alteration and reduction of microbial colonization in early life [110]. This posits that modern westernized changes in environmental exposures impact the developing microbiota composition, resulting in immune responses to allergens and an increase in allergic and immune-mediated diseases. Observational and molecular epidemiology cohort studies have demonstrated that alterations in infant gut microbiotas are associated with immune-mediated health outcomes such as allergy, atopy, and asthma [73,99–105,110]. Allergic diseases are, by definition, immunological pathologies, and multiple investigations have described associations between dysbiosis in a critical window of immune development and immune-mediated diseases, which often persist into adulthood [66,99,105,109,110].

Low bacterial diversity and paucity of specific microbes in infancy, specifically *Bifidobacterium*, *Akkermansia*, and *Faecalibacterium*, high abundances of fungi, and distinct fecal metabolites are associated with higher risk of atopy and asthma [105]. Specific microbial patterns are associated with infants with food allergies, including paucity of *Bacteroides* and *Clostridium*, and another cohort study identified lower diversity at age 3 months with increased Enterobacteriaceae, and higher abundances of Bacteroides at 1 year of age [100,109]. Dysbiosis is a disruption in the composition of the microbiota in its developmental trajectory, and often represents less diversity of microbes with altered ratios of immune-mediating taxa. Dysbiosis is associated with multiple lifelong diseases, particularly if this disruption occurs during the developmental phase when immune training is occurring [118]. Transplantation of the microbiome in numerous disease states to germ-free animals results in phenotypic expression of that disease, highlighting opportunities for research for targeted interventions [119,120].

Harnessing the microbiome-immune training critical window for vaccine optimization

Immune maturation drives responses to invading pathogens and antigens including vaccines. Several studies have demonstrated the effect of the developing infant microbiome on vaccine responses [121–126]. Specifically, the relative abundance of Actinobacteria including *Bifidobacterium* was positively correlated with IgG and T-cell responses to Bacillus Calmette–Guérin (BCG), oral polio, and tetanus toxoid vaccinations in Bangladesh infants, suggesting that vaccine responsiveness could be improved by promoting dietary *Bifidobacteria* and minimizing dysbiosis early in infancy [124]. The intestinal microbiota is postulated to impact oral vaccine efficacy via suppression of IgA and IgG [127]. The gut microbiota may prime mucosal antigen-specific B cell responses [128] and may influence vaccine efficacy. Germ-free mouse models have clarified the key role microbes play in vaccine response, in TLR5 −/− mice, with microbiota restoration resulting in competent vaccine-specific IgG response [125]. The relationship between the developing microbiome, helper T-cell ratios of Th1 and Th2 responses, Treg cells, NK cells, and IgG subclass composition is being examined more closely to clarify the time period during which microbiota-driven modulation of infant vaccine response can be explored [18,114]. Specific commensals, such as *Bifidobacterium*, are associated with enhanced oral vaccine efficacy [74].

Sex differences and microbiome immune mediation

The innate and adaptive immune system are known to have sex-specific differences, including higher numbers of adaptive immune cells such as T helper cells, B cells and higher production of IL-2 producing lymphocytes in females; males have higher frequency of NK cells and monocytes [129,130]. Sex differences in immunity are postulated to relate to sex hormones and genetic immune responses [20,130]. The intestinal tract and its GALT impact both local and systemic immunity and inflammation, regulating immune homeostasis in diverse organ systems [131,132]. The microbiota appears to differ between the sexes in animal models and in humans [133–135], and these differences are associated with sex-linked differences in inflammation both in the GALT, systemic inflammation, and inflammatory diseases such as systemic lupus erythematosus and type 1 diabetes [129,130,136]. Murine models have enabled demonstration of enhanced intestinal innate immunity in males and reduced adaptive immunity when compared with females [137]; in rats lower proportions of T cells and macrophages were found in the mesenteric lymph nodes in males [138]. A transfer study in germ-free male and female mice (where female microbiota was transplanted into germ-free male mice and vice versa) resulted in sex-dependent differences in immune function, including higher percentages of Foxp3 + cells in the Peyer's patches and mesenteric lymph nodes of female recipients of male microbiota [139]. Transplantation of male murine microbiota to female nonobese diabetic mice provided protection against development of type one diabetes [140,141]. The dimorphic differences in both the microbiome [141−143] and innate and adaptive immunity [137] identified demand that investigations of sex differences in the microbiota and its impact on immunity be considered as interventions for manipulating the microbiota to prevent or treat disease or optimize health are strategized.

Conclusions and relevance across the life course

The newborn microbiome develops in parallel with innate and adaptive immunity in the first months of life, demonstrating a complex bidirectional interplay required for maturation and competence. The interdependent microbe-driven immune system is profoundly influenced by maternal breast milk through multiple bioactive factors including prebiotics and maternal milk probiotics that impact nutrient metabolism, vitamin formation and absorption, and production of metabolites that directly enhance maturation and generation of immune cells. Epidemiological studies in humans and their associated investigations in animal models have clarified mechanisms behind microbe-driven immune maturation, and developmental aberrations result in immune-mediated diseases such as diabetes impacting children and adults. A critical window of time during this coevolution is becoming clarified, affording opportunities to intervene and reverse disease trajectory or prevent diseases such as asthma and type 1 diabetes, which persist into adulthood as lifelong diseases. Similarly, vaccine efficacy that is impacted by microbiome-influenced immune competence provides opportunities for interventions to optimize disease prevention. Sex differences in immunity and autoimmune diseases are becoming elucidated, with animal models clarifying that sex-specific differences in the microbiome may be driving immune-mediated health outcomes. Investigation of the role of the developing infant intestinal microbiome in a critical window during which the immune system matures provides opportunities to leverage novel interventions to optimize health for a lifetime.

References

[1] Tsafaras GP, Ntontsi P, Xanthou G. Advantages and limitations of the neonatal immune sSystem. Front Pediatr 2020;8:5. Available from: https://doi.org/10.3389/fped.2020.00005 PubMed PMID: 32047730; PMCID: PMC6997472.

[2] Belderbos M, Levy O, Bont L. Neonatal innate immunity in allergy development. Curr Opin Pediatr 2009;21(6):762−9. Available from: https://doi.org/10.1097/MOP.0b013e3283325e3a PubMed PMID: 19770766.

[3] Levy O. Innate immunity of the newborn: basic mechanisms and clinical correlates. Nat Rev Immunol 2007;7(5):379–90. Available from: https://doi.org/10.1038/nri2075 PubMed PMID: 17457344.
[4] PrabhuDas M, Adkins B, Gans H, King C, Levy O, Ramilo O, et al. Challenges in infant immunity: implications for responses to infection and vaccines. Nat Immunol 2011;12 (3):189–94. Available from: https://doi.org/10.1038/ni0311-189 PubMed PMID: 21321588.
[5] Ygberg S, Nilsson A. The developing immune system—from foetus to toddler. Acta Paediatr 2012;101(2):120–7. Available from: https://doi.org/10.1111/j.1651-2227.2011.02494.x PubMed PMID: 22003882.
[6] Kollmann TR, Kampmann B, Mazmanian SK, Marchant A, Levy O. Protecting the newborn and young infant from infectious diseases: lessons from immune ontogeny. Immunity 2017;46(3):350–63. Available from: https://doi.org/10.1016/j.immuni.2017.03.009 Epub 2017/03/23. PubMed PMID: 28329702.
[7] Dowling DJ, Levy O. Ontogeny of early life immunity. Trends Immunol 2014;35 (7):299–310. Available from: https://doi.org/10.1016/j.it.2014.04.007 PubMed PMID: 24880460; PMCID: PMC4109609.
[8] Cho I, Blaser MJ. The human microbiome: at the interface of health and disease. Nat Rev Genet 2012;13(4):260–70. Available from: https://doi.org/10.1038/nrg3182 Epub 2012/03/14. PubMed PMID: 22411464; PMCID: PMC3418802.
[9] de Muinck EJ, Trosvik P. Individuality and convergence of the infant gut microbiota during the first year of life. Nat Commun 2018;9(1):2233. Available from: https://doi.org/10.1038/s41467-018-04641-7 Epub 2018/06/10. PubMed PMID: 29884786; PMCID: PMC5993781.
[10] Gomez de Aguero M, Ganal-Vonarburg SC, Fuhrer T, Rupp S, Uchimura Y, Li H, et al. The maternal microbiota drives early postnatal innate immune development. Science 2016;351(6279):1296–302. Available from: https://doi.org/10.1126/science.aad2571 PubMed PMID: 26989247.
[11] Nyangahu DD, Lennard KS, Brown BP, Darby MG, Wendoh JM, Havyarimana E, et al. Disruption of maternal gut microbiota during gestation alters offspring microbiota and immunity. Microbiome 2018;6(1):124. Available from: https://doi.org/10.1186/s40168-018-0511-7 PubMed PMID:29981583; PMCID: PMC6035804.
[12] Hooper LV, Littman DR, Macpherson AJ. Interactions between the microbiota and the immune system. Science 2012;336(6086):1268–73. Available from: https://doi.org/10.1126/science.1223490 Epub 2012/06/08. PubMed PMID: 22674334; PMCID: PMC4420145.
[13] Maynard CL, Elson CO, Hatton RD, Weaver CT. Reciprocal interactions of the intestinal microbiota and immune system. Nature 2012;489(7415):231–41. Available from: https://doi.org/10.1038/nature11551 Epub 2012/09/14. PubMed PMID: 22972296; PMCID: PMC4492337.
[14] Round JL, Mazmanian SK. The gut microbiota shapes intestinal immune responses during health and disease. Nat Rev Immunol 2009;9(5):313–23. Available from: https://doi.org/10.1038/nri2515 Epub 2009/04/04. PubMed PMID: 19343057; PMCID: PMC4095778.
[15] Philbin VJ, Levy O. Developmental biology of the innate immune response: implications for neonatal and infant vaccine development. Pediatr Res 2009;65(5 Pt 2):98R–105R. Available from: https://doi.org/10.1203/PDR.0b013e31819f195d PubMed PMID: 19918215; PMCID: PMC2795575.
[16] Hooper LV, Macpherson AJ. Immune adaptations that maintain homeostasis with the intestinal microbiota. Nat Rev Immunol 2010;10(3):159–69. Available from: https://doi.org/10.1038/nri2710 Epub 2010/02/26. PubMed PMID: 20182457.

[17] Filias A, Theodorou GL, Mouzopoulou S, Varvarigou AA, Mantagos S, Karakantza M. Phagocytic ability of neutrophils and monocytes in neonates. BMC Pediatr 2011;11:29. Available from: https://doi.org/10.1186/1471-2431-11-29 Epub 2011/04/14. PubMed PMID: 21492472; PMCID: PMC3094219.

[18] Simon AK, Hollander GA, McMichael A. Evolution of the immune system in humans from infancy to old age. Proc Biol Sci 2015;282(1821):20143085. Available from: https://doi.org/10.1098/rspb.2014.3085 Epub 2015/12/25. PubMed PMID: 26702035; PMCID: PMC4707740.

[19] Rios-Covian D, Ruas-Madiedo P, Margolles A, Gueimonde M, de Los Reyes-Gavilan CG, Salazar N. Intestinal short chain fatty acids and their link with diet and human health. Front in Microbiol 2016;7:185. Available from: https://doi.org/10.3389/fmicb.2016.00185 PubMed PMID: 26925050; PMCID: PMC4756104.

[20] Vemuri R, Sylvia KE, Klein SL, Forster SC, Plebanski M, Eri R, et al. The microgenderome revealed: sex differences in bidirectional interactions between the microbiota, hormones, immunity and disease susceptibility. Semimmunopathology 2019;41(2):265–75. Available from: https://doi.org/10.1007/s00281-018-0716-7 Epub 2018/10/10. PubMed PMID: 30298433; PMCID: PMC6500089.

[21] Kim YS, Unno T, Kim BY, Park MS. Sex differences in gut microbiota. World J Mens Health 2020;38(1):48–60. Available from: https://doi.org/10.5534/wjmh.190009 Epub 2019/04/01. PubMed PMID: 30929328; PMCID: PMC6920072.

[22] Huang XZ, Zhu LB, Li ZR, Lin J. Bacterial colonization and intestinal mucosal barrier development. World J Clin Pediatr 2013;2(4):46–53. Available from: https://doi.org/10.5409/wjcp.v2.i4.46 PubMed PMID: 25254174; PMCID: PMC4145654.

[23] Brestoff JR, Artis D. Commensal bacteria at the interface of host metabolism and the immune system. Nat Immunol 2013;14(7):676–84. Available from: https://doi.org/10.1038/ni.2640 Epub 2013/06/20. PubMed PMID: 23778795; PMCID: PMC4013146.

[24] Yu JC, Khodadadi H, Malik A, Davidson B, Salles E, Bhatia J, et al. Innate immunity of neonates and infants. Front Immunol 2018;9:1759. Available from: https://doi.org/10.3389/fimmu.2018.01759 Epub 2018/08/15. PubMed PMID: 30105028; PMCID: PMC6077196.

[25] Basha S, Surendran N, Pichichero M. Immune responses in neonates. Expert Rev Clin Immunol 2014;10(9):1171–84. Available from: https://doi.org/10.1586/1744666x.2014.942288 PubMed PMID: WOS:000341522900008.

[26] Tregoning JS, Wang BL, McDonald JU, Yamaguchi Y, Harker JA, Goritzka M, et al. Neonatal antibody responses are attenuated by interferon-gamma produced by NK and T cells during RSV infection. Proc Natl Acad Sci U S A 2013;110(14):5576–81. Available from: https://doi.org/10.1073/pnas.1214247110 Epub 2013/03/20. PubMed PMID: 23509276; PMCID: PMC3619373.

[27] Jakaitis BM, Denning PW. Human breast milk and the gastrointestinal innate immune system. Clin Perinatol 2014;41(2):423–35. Available from: https://doi.org/10.1016/j.clp.2014.02.011 Epub 2014/05/31. PubMed PMID: 24873841; PMCID: PMC4414019.

[28] Mowat AM. Anatomical basis of tolerance and immunity to intestinal antigens. Nat Rev Immunol 2003;3(4):331–41. Available from: https://doi.org/10.1038/nri1057 Epub 2003/04/02. PubMed PMID: 12669023.

[29] Cerovic V, Bain CC, Mowat AM, Milling SW. Intestinal macrophages and dendritic cells: what's the difference? Trends Immunol 2014;35(6):270–7. Available from: https://doi.org/10.1016/j.it.2014.04.003 Epub 2014/05/06. PubMed PMID: 24794393.

[30] Chistiakov DA, Bobryshev YV, Kozarov E, Sobenin IA, Orekhov AN. Intestinal mucosal tolerance and impact of gut microbiota to mucosal tolerance. Front Microbiol 2014;5:781.

Available from: https://doi.org/10.3389/fmicb.2014.00781 PubMed PMID: 25628617; PMCID: PMC4292724.

[31] Donnet-Hughes A, Perez PF, Dore J, Leclerc M, Levenez F, Benyacoub J, et al. Potential role of the intestinal microbiota of the mother in neonatal immune education. Proc Nutr Soc 2010;69(3):407–15. Available from: https://doi.org/10.1017/S0029665110001898 Epub 2010/07/17. PubMed PMID: 20633308.

[32] Underwood MA, Gilbert WM, Sherman MP. Amniotic fluid: not just fetal urine anymore. J Perinatol 2005;25(5):341–8. Available from: https://doi.org/10.1038/sj.jp.7211290 PubMed PMID: 15861199.

[33] Mold JE, Michaelsson J, Burt TD, Muench MO, Beckerman KP, Busch MP, et al. Maternal alloantigens promote the development of tolerogenic fetal regulatory T cells in utero. Science 2008;322(5907):1562–5. Available from: https://doi.org/10.1126/science.1164511 PubMed PMID:19056990; PMCID: PMC2648820.

[34] Basha S, Surendran N, Pichichero M. Immune responses in neonates. Expert Rev Clin Immunol 2014;10(9):1171–84. Available from: https://doi.org/10.1586/1744666x.2014.942288 Epub 2014/08/05. PubMed PMID: 25088080; PMCID: PMC4407563.

[35] van den Berg JP, Westerbeek EA, van der Klis FR, Berbers GA, van Elburg RM. Transplacental transport of IgG antibodies to preterm infants: a review of the literature. Early Hum Dev 2011;87(2):67–72. Available from: https://doi.org/10.1016/j.earlhumdev.2010.11.003 PubMed PMID: 21123010.

[36] Coker MO, Hoen AG, Dade E, Lundgren S, Li Z, Wong AD, et al. Specific class of intrapartum antibiotics relates to maturation of the infant gut microbiota: a prospective cohort study. BJOG 2020;127(2):217–27. Available from: https://doi.org/10.1111/1471-0528.15799 Epub 2019/06/05. PubMed PMID: 31006170; PMCID: PMC6803026.

[37] Macpherson AJ, de Aguero MG, Ganal-Vonarburg SC. How nutrition and the maternal microbiota shape the neonatal immune system. Nat Rev Immunol 2017;17(8):508–17. Available from: https://doi.org/10.1038/nri.2017.58 Epub 2017/06/13. PubMed PMID: 28604736.

[38] Perez PF, Doré J, Leclerc M, Levenez F, Benyacoub J, Serrant P, et al. Bacterial imprinting of the neonatal immune system: lessons from maternal cells. Pediatrics 2007;119(3): e724–32. Available from: https://doi.org/10.1542/peds.2006-1649 PubMed PMID: 17332189.

[39] Penders J, Thijs C, Vink C, Stelma FF, Snijders B, Kummeling I, et al. Factors influencing the composition of the intestinal microbiota in early infancy. Pediatrics 2006;118 (2):511–5121. Available from: https://doi.org/10.1542/peds.2005-2824 Epub 2006/08/03. doi: 118/2/511 [pii]. PubMed PMID: 16882802.

[40] Madan JC, Hoen AG, Lundgren SN, Farzan SF, Cottingham KL, Morrison HG, et al. Association of cesarean delivery and formula supplementation with the intestinal microbiome of 6-week-old infants. JAMA Pediatr 2016;170(3):212–19. Available from: https://doi.org/10.1001/jamapediatrics.2015.3732 Epub 2016/01/12. PubMed PMID: 26752321; PMCID: PMC4783194.

[41] Singh SB, Madan J, Coker M, Hoen A, Baker ER, Karagas MR, et al. Does birth mode modify associations of maternal pre-pregnancy BMI and gestational weight gain with the infant gut microbiome? Int J Obes (2005) 2020;44(1):23–32. Available from: https://doi.org/10.1038/s41366-018-0273-0 PubMed PMID: 30765892; PMCID: PMC6694002.

[42] McGuire MK, McGuire MA. Human milk: mother nature's prototypical probiotic food? Adv Nutr 2015;6(1):112–23. Available from: https://doi.org/10.3945/an.114.007435 Epub 2015/01/17. PubMed PMID: 25593150; PMCID: PMC4288270.

[43] Iwasaki A, Medzhitov R. Control of adaptive immunity by the innate immune system. Nat Immunol 2015;16(4):343–53. Available from: https://doi.org/10.1038/ni.3123 Epub 2015/03/20. PubMed PMID: 25789684; PMCID: PMC4507498.

[44] Newburg DS, Walker WA. Protection of the neonate by the innate immune system of developing gut and of human milk. Pediatric Res 2007;61(1):2–8. Available from: https://doi.org/10.1203/01.pdr.0000250274.68571.18 Epub 2007/01/11. PubMed PMID: 17211132.

[45] Cacho NT, Lawrence RM. Innate immunity and breast milk. Front Immunol 2017;8:584. Available from: https://doi.org/10.3389/fimmu.2017.00584 Epub 2017/06/15. PubMed PMID: 28611768; PMCID: PMC5447027.

[46] Jarvinen KM, Suomalainen H. Leucocytes in human milk and lymphocyte subsets in cow's milk-allergic infants. Pediatr Allergy Immunol 2002;13(4):243–54. Available from: https://doi.org/10.1034/j.1399-3038.2002.00087.x Epub 2002/10/23. PubMed PMID: 12390440.

[47] Cabinian A, Sinsimer D, Tang M, Zumba O, Mehta H, Toma A, et al. Transfer of maternal immune cells by breastfeeding: maternal cytotoxic T lymphocytes present in breast milk localize in the Peyer's Patches of the nursed infant. PLoS One 2016;11(6):e0156762. Available from: https://doi.org/10.1371/journal.pone.0156762 Epub 2016/06/11. PubMed PMID: 27285085; PMCID: PMC4902239.

[48] He Y, Liu S, Kling DE, Leone S, Lawlor NT, Huang Y, et al. The human milk oligosaccharide 2′-fucosyllactose modulates CD14 expression in human enterocytes, thereby attenuating LPS-induced inflammation. Gut 2016;65(1):33–46. Available from: https://doi.org/10.1136/gutjnl-2014-307544 Epub 2014/11/29. PubMed PMID: 25431457.

[49] He W, Rahimnejad S, Wang L, Song K, Lu K, Zhang C. Effects of organic acids and essential oils blend on growth, gut microbiota, immune response and disease resistance of Pacific white shrimp (Litopenaeus vannamei) against *Vibrio parahaemolyticus*. Fish Shellfish Immunol 2017;70:164–73. Available from: https://doi.org/10.1016/j.fsi.2017.09.007 PubMed PMID: 28882791.

[50] Murphy K, Curley D, O'Callaghan TF, O'Shea CA, Dempsey EM, O'Toole PW, et al. The composition of human milk and infant faecal microbiota over the first three months of life: a pilot study. Sci Rep 2017;7:40597. Available from: https://doi.org/10.1038/srep40597 PubMed PMID: 28094284; PMCID: PMC5240090.

[51] Lundgren SN, Madan JC, Emond JA, Morrison HG, Christensen BC, Karagas MR, et al. Maternal diet during pregnancy is related with the infant stool microbiome in a delivery mode-dependent manner. Microbiome 2018;6(1):109. Available from: https://doi.org/10.1186/s40168-018-0490-8 Epub 2018/07/06. PubMed PMID: 29973274; PMCID: PMC6033232.

[52] Lackey KA, Williams JE, Meehan CL, Zachek JA, Benda ED, Price WJ, et al. What's normal? Microbiomes in human milk and infant feces are related to each other but vary geographically: the INSPIRE study. Front Nutr 2019;6:45. Available from: https://doi.org/10.3389/fnut.2019.00045 Epub 2019/05/07. PubMed PMID: 31058158; PMCID: PMC6479015.

[53] Gomez-Gallego C, Garcia-Mantrana I, Salminen S, Collado MC. The human milk microbiome and factors influencing its composition and activity. Semin Fetal Neonatal Med 2016;21(6):400–5. Available from: https://doi.org/10.1016/j.siny.2016.05.003 Epub 2016/06/12. PubMed PMID: 27286644.

[54] Fitzstevens JL, Smith KC, Hagadorn JI, Caimano MJ, Matson AP, Brownell EA. Systematic review of the human milk microbiota. Nutr Clin Pract 2017;32(3):354–64.

Available from: https://doi.org/10.1177/0884533616670150 Epub 2016/09/30. PubMed PMID: 27679525.

[55] Newburg DS, He Y. Neonatal gut microbiota and human milk hlycans cooperate to attenuate infection and inflammation. Clin Obstet Gynecol 2015;58(4):814–26. Available from: https://doi.org/10.1097/grf.0000000000000156 Epub 2015/10/13. PubMed PMID: 26457857.

[56] Williams JE, Carrothers JM, Lackey KA, Beatty NF, Brooker SL, Peterson HK, et al. Strong multivariate relations exist among milk, oral, and fecal microbiomes in mother-infant dyads during the first six months postpartum. J Nutr 2019;149(6):902–14. Available from: https://doi.org/10.1093/jn/nxy299 Epub 2019/05/08. PubMed PMID: 31063198; PMCID: PMC6543206.

[57] Marcobal A, Barboza M, Froehlich JW, Block DE, German JB, Lebrilla CB, et al. Consumption of human milk oligosaccharides by gut-related microbes. J Agric Food Chem 2010;58(9):5334–40. Available from: https://doi.org/10.1021/jf9044205 PubMed PMID: 20394371; PMCID: PMC2866150.

[58] Marcobal A, Sonnenburg JL. Human milk oligosaccharide consumption by intestinal microbiota. Clin Microbiol Infect 2012;18(Suppl 4):12–15. Available from: https://doi.org/10.1111/j.1469-0691.2012.03863.x PubMed PMID: 22647041; PMCID: PMC3671919.

[59] Ardeshir A, Narayan NR, Mendez-Lagares G, Lu D, Rauch M, Huang Y, et al. Breast-fed and bottle-fed infant rhesus macaques develop distinct gut microbiotas and immune systems. Sci Transl Med 2014;6(252). Available from: https://doi.org/10.1126/scitranslmed.3008791 252ra120. Epub 2014/09/05. PubMed PMID: 25186175; PMCID: PMC4362692.

[60] Georgountzou A, Papadopoulos NG. Postnatal innate immune development: from birth to adulthood. Front Immunol 2017;8:957. Available from: https://doi.org/10.3389/fimmu.2017.00957 Epub 2017/08/30. PubMed PMID: 28848557; PMCID: PMC5554489.

[61] Cormick G, Betran AP, Harbron J, Dannemann Purnat T, Parker C, Hall D, et al. Calcium, pre-eclampsia study G. Are women with history of pre-eclampsia starting a new pregnancy in good nutritional status in South Africa and Zimbabwe? BMC Pregnancy Childbirth 2018;18(1):236. Available from: https://doi.org/10.1186/s12884-018-1885-z PubMed PMID: 29907146; PMCID: PMC6003186.

[62] Levy O. Innate immunity of the human newborn: distinct cytokine responses to LPS and other toll-like receptor agonists. J Endotoxin Res 2005;11(2):113–16. Available from: https://doi.org/10.1179/096805105X37376 PubMed PMID: 15949138.

[63] Angelone DF, Wessels MR, Coughlin M, Suter EE, Valentini P, Kalish LA, et al. Innate immunity of the human newborn is polarized toward a high ratio of IL-6/TNF-alpha production in vitro and in vivo. Pediatr Res 2006;60(2):205–9. Available from: https://doi.org/10.1203/01.pdr.0000228319.10481.ea PubMed PMID: 16864705.

[64] Akira S. Toll-like receptors and innate immunity. Adv Immunol 2001;78:1–56. Available from: https://doi.org/10.1016/s0065-2776(01)78001-7 PubMed PMID: 11432202.

[65] Gensollen T, Iyer SS, Kasper DL, Blumberg RS. How colonization by microbiota in early life shapes the immune system. Science 2016;352(6285):539–44. Available from: https://doi.org/10.1126/science.aad9378 Epub 2016/04/30. PubMed PMID: 27126036; PMCID: PMC5050524.

[66] Ohnmacht C. Microbiota, regulatory T cell subsets, and allergic disorders. Allergo J Int 2016;25(5):114–23. Available from: https://doi.org/10.1007/s40629-016-0118-0 Epub 2016/09/23. PubMed PMID: 27656354; PMCID: PMC5016534.

[67] Adkins B, Leclerc C, Marshall-Clarke S. Neonatal adaptive immunity comes of age. Nat Rev Immunol 2004;4(7):553–64. Available from: https://doi.org/10.1038/nri1394 PubMed PMID: 15229474.

[68] Janzon A, Goodrich JK, Koren O, Group TS, Waters JL, Ley RE. Interactions between the gut microbiome and mucosal immunoglobulins A, M, and G in the developing infant gut. mSystems 2019;4(6). Available from: https://doi.org/10.1128/mSystems.00612-19 PubMed PMID: 31771976; PMCID: PMC6880043.

[69] Perkkio M, Savilahti E. Time of appearance of immunoglobulin-containing cells in the mucosa of the neonatal intestine. Pediatr Res 1980;14(8):953–5. Available from: https://doi.org/10.1203/00006450-198008000-00012 PubMed PMID: 7191555.

[70] Knox WF. Restricted feeding and human intestinal plasma cell development. Arch Dis Child 1986;61(8):744–9. Available from: https://doi.org/10.1136/adc.61.8.744 PubMed PMID: 3090947; PMCID: PMC1777918.

[71] Macpherson AJ, Harris NL. Interactions between commensal intestinal bacteria and the immune system. Nat Rev Immunol 2004;4(6):478–85. Available from: https://doi.org/10.1038/nri1373 Epub 2004/06/03. PubMed PMID: 15173836.

[72] Atarashi K, Nishimura J, Shima T, Umesaki Y, Yamamoto M, Onoue M, et al. ATP drives lamina propria T(H)17 cell differentiation. Nature 2008;455(7214):808–12. Available from: https://doi.org/10.1038/nature07240 Epub 2008/08/22. PubMed PMID: 18716618.

[73] Belkaid Y, Hand TW. Role of the microbiota in immunity and inflammation. Cell 2014;157(1):121–41. Available from: https://doi.org/10.1016/j.cell.2014.03.011 Epub 2014/04/01. PubMed PMID: 24679531; PMCID: PMC4056765.

[74] Sjogren YM, Tomicic S, Lundberg A, Bottcher MF, Bjorksten B, Sverremark-Ekstrom E, et al. Influence of early gut microbiota on the maturation of childhood mucosal and systemic immune responses. Clin Exp Allergy 2009;39(12):1842–51. Available from: https://doi.org/10.1111/j.1365-2222.2009.03326.x Epub 2009/09/09. doi: CEA3326 [pii] PubMed PMID: 19735274.

[75] Zheng S, Zhao T, Yuan S, Yang L, Ding J, Cui L, et al. Immunodeficiency promotes adaptive alterations of host gut microbiome: an observational metagenomic study in mice. Front Microbiol 2019;10:2415. Available from: https://doi.org/10.3389/fmicb.2019.02415 PubMed PMID: 31781050; PMCID: PMC6853035.

[76] Fouhse JM, Yang K, More-Bayona J, Gao Y, Goruk S, Plastow G, et al. Neonatal exposure to amoxicillin alters long-term immune response despite transient effects on gut-microbiota in piglets. Front Immunol 2019;10:2059. Available from: https://doi.org/10.3389/fimmu.2019.02059 PubMed PMID: 31552023; PMCID: PMC6737505.

[77] Kawamoto S, Maruya M, Kato LM, Suda W, Atarashi K, Doi Y, et al. Foxp3(+) T cells regulate immunoglobulin a selection and facilitate diversification of bacterial species responsible for immune homeostasis. Immunity 2014;41(1):152–65. Available from: https://doi.org/10.1016/j.immuni.2014.05.016 Epub 2014/07/10. PubMed PMID: 25017466.

[78] Stiemsma LT, Turvey SE. Asthma and the microbiome: defining the critical window in early life. Allergy Asthma Clin Immunol 2017;13:3. Available from: https://doi.org/10.1186/s13223-016-0173-6 Epub 2017/01/13. PubMed PMID: 28077947; PMCID: PMC5217603.

[79] Bisgaard H, Li N, Bonnelykke K, Chawes BL, Skov T, Paludan-Muller G, et al. Reduced diversity of the intestinal microbiota during infancy is associated with increased risk of allergic disease at school age. J Allergy Clin Immunol 2011;128(3). Available from: https://doi.org/10.1016/j.jaci.2011.04.060 646-52 e1-5. Epub 2011/07/26. PubMed PMID: 21782228.

[80] Smith K, McCoy KD, Macpherson AJ. Use of axenic animals in studying the adaptation of mammals to their commensal intestinal microbiota. Semin Immunol 2007;19(2):59–69. Available from: https://doi.org/10.1016/j.smim.2006.10.002 PubMed PMID: 17118672.
[81] Falk PG, Hooper LV, Midtvedt T, Gordon JI. Creating and maintaining the gastrointestinal ecosystem: what we know and need to know from gnotobiology. Microbiol Mol Biol Rev 1998;62(4):1157–70 PubMed PMID: 9841668; PMCID: PMC98942.
[82] Pollard M, Sharon N. Responses of the Peyer's patches in germ-free mice to antigenic stimulation. Infect Immun 1970;2(1):96–100 PubMed PMID: 16557807; PMCID: PMC415970.
[83] Sommer F, Backhed F. The gut microbiota—masters of host development and physiology. Nat Rev Microbiol 2013;11(4):227–38. Available from: https://doi.org/10.1038/nrmicro2974 PubMed PMID: 23435359.
[84] Sudo N, Sawamura S, Tanaka K, Aiba Y, Kubo C, Koga Y. The requirement of intestinal bacterial flora for the development of an IgE production system fully susceptible to oral tolerance induction. J Immunol 1997;159(4):1739–45 Epub 1997/08/15. PubMed PMID: 9257835.
[85] Butler JE, Weber P, Sinkora M, Baker D, Schoenherr A, Mayer B, et al. Antibody repertoire development in fetal and neonatal piglets. VIII. Colonization is required for newborn piglets to make serum antibodies to T-dependent and type 2 T-independent antigens. J Immunol 2002;169(12):6822–30. Available from: https://doi.org/10.4049/jimmunol.169.12.6822 PubMed PMID: 12471114.
[86] Ibnou-Zekri N, Blum S, Schiffrin EJ, von der Weid T. Divergent patterns of colonization and immune response elicited from two intestinal Lactobacillus strains that display similar properties in vitro. Infect Immun 2003;71(1):428–36. Available from: https://doi.org/10.1128/iai.71.1.428-436.2003 PubMed PMID: 12496193; PMCID: PMC143181.
[87] Obata Y, Furusawa Y, Endo TA, Sharif J, Takahashi D, Atarashi K, et al. The epigenetic regulator Uhrf1 facilitates the proliferation and maturation of colonic regulatory T cells. Nat Immunol 2014;15(6):571–9. Available from: https://doi.org/10.1038/ni.2886 Epub 2014/04/28. PubMed PMID: 24777532.
[88] Ivanov II, Atarashi K, Manel N, Brodie EL, Shima T, Karaoz U, et al. Induction of intestinal Th17 cells by segmented filamentous bacteria. Cell 2009;139(3):485–98. Available from: https://doi.org/10.1016/j.cell.2009.09.033 Epub 2009/10/20. PubMed PMID: 19836068; PMCID: 2796826.
[89] Grönlund MM, Arvilommi H, Kero P, Lehtonen OP, Isolauri E. Importance of intestinal colonisation in the maturation of humoral immunity in early infancy: a prospective follow up study of healthy infants aged 0-6 months. Arch Dis Child Fetal Neonatal Ed 2000;83 (3):F186–92. Available from: https://doi.org/10.1136/fn.83.3.f186 PubMed PMID: 11040166; PMCID: PMC1721174.
[90] Marschan E, Kuitunen M, Kukkonen K, Poussa T, Sarnesto A, Haahtela T, et al. Probiotics in infancy induce protective immune profiles that are characteristic for chronic low-grade inflammation. Clin Exp Allergy 2008;38(4):611–18. Available from: https://doi.org/10.1111/j.1365-2222.2008.02942.x Epub 2008/02/11. PubMed PMID: 18266878.
[91] Pellicciotta M, Rigoni R, Falcone EL, Holland SM, Villa A, Cassani B. The microbiome and immunodeficiencies: lessons from rare diseases. J Autoimmun 2019;98:132–48. Available from: https://doi.org/10.1016/j.jaut.2019.01.008 Epub 2019/02/02. PubMed PMID: 30704941.
[92] Furusawa Y, Obata Y, Fukuda S, Endo TA, Nakato G, Takahashi D, et al. Commensal microbe-derived butyrate induces the differentiation of colonic regulatory T cells. Nature 2013;504(7480):446–50. Available from: https://doi.org/10.1038/nature12721 PubMed PMID: 24226770.

[93] Chang PV, Hao L, Offermanns S, Medzhitov R. The microbial metabolite butyrate regulates intestinal macrophage function via histone deacetylase inhibition. Proc Natl Acad Sci U S A 2014;111(6):2247–52. Available from: https://doi.org/10.1073/pnas.1322269111 PubMed PMID: 24390544; PMCID: PMC3926023.

[94] Zeevi D, Korem T, Godneva A, Bar N, Kurilshikov A, Lotan-Pompan M, et al. Structural variation in the gut microbiome associates with host health. Nature 2019;568 (7750):43–8. Available from: https://doi.org/10.1038/s41586-019-1065-y PubMed PMID: 30918406.

[95] Ozkan H, Topsakal S, Ozmen O. Investigation of the diabetic effects of maternal high-glucose diet on rats. Biomed Pharmacother 2019;110:609–17. Available from: https://doi.org/10.1016/j.biopha.2018.12.011 PubMed PMID: 30537678.

[96] Mazmanian SK, Liu CH, Tzianabos AO, Kasper DL. An immunomodulatory molecule of symbiotic bacteria directs maturation of the host immune system. Cell 2005;122 (1):107–18. Available from: https://doi.org/10.1016/j.cell.2005.05.007 PubMed PMID: 16009137.

[97] Weintraub A. Immunology of bacterial polysaccharide antigens. Carbohydr Res 2003;338(23):2539–47. Available from: https://doi.org/10.1016/j.carres.2003.07.008 PubMed PMID: 14670715.

[98] Ramakrishna C, Kujawski M, Chu H, Li L, Mazmanian SK, Cantin EM. Bacteroides fragilis polysaccharide A induces IL-10 secreting B and T cells that prevent viral encephalitis. Nat Commun 2019;10(1):2153. Available from: https://doi.org/10.1038/s41467-019-09884-6 PubMed PMID: 31089128; PMCID: PMC6517419.

[99] Arrieta MC, Stiemsma LT, Dimitriu PA, Thorson L, Russell S, Yurist-Doutsch S, et al. Early infancy microbial and metabolic alterations affect risk of childhood asthma. Sci Transl Med 2015;7(307). Available from: https://doi.org/10.1126/scitranslmed.aab2271 307ra152. Epub 2015/10/02. PubMed PMID: 26424567.

[100] Azad MB, Konya T, Guttman DS, Field CJ, Sears MR, HayGlass KT, et al. Infant gut microbiota and food sensitization: associations in the first year of life. Clin Exp Allergy 2015;45(3):632–43. Available from: https://doi.org/10.1111/cea.12487 Epub 2015/01/21. PubMed PMID: 25599982.

[101] Borbet TC, Zhang X, Muller A, Blaser MJ. The role of the changing human microbiome in the asthma pandemic. J Allergy Clin Immunol 2019;144(6):1457–66. Available from: https://doi.org/10.1016/j.jaci.2019.10.022 Epub 2019/12/10. PubMed PMID: 31812180; PMCID: PMC6905199.

[102] Butel MJ, Waligora-Dupriet AJ, Wydau-Dematteis S. The developing gut microbiota and its consequences for health. J Dev Orig Health Dis 2018;9(6):590–7. Available from: https://doi.org/10.1017/s2040174418000119 Epub 2018/03/23. PubMed PMID: 29562949.

[103] Dang AT, Marsland BJ. Microbes, metabolites, and the gut-lung axis. Mucosal Immunol 2019;12(4):843–50. Available from: https://doi.org/10.1038/s41385-019-0160-6 Epub 2019/04/13. PubMed PMID: 30976087.

[104] Felix KM, Tahsin S, Wu HJ. Host-microbiota interplay in mediating immune disorders. Ann N Y Acad Sci 2018;1417(1):57–70. Available from: https://doi.org/10.1111/nyas.13508 Epub 2017/10/07. PubMed PMID: 28984367; PMCID: PMC5889363.

[105] Fujimura KE, Sitarik AR, Havstad S, Lin DL, Levan S, Fadrosh D, et al. Neonatal gut microbiota associates with childhood multisensitized atopy and T cell differentiation. Nat Med 2016;22(10):1187–91. Available from: https://doi.org/10.1038/nm.4176 Epub 2016/09/13. PubMed PMID: 27618652; PMCID: PMC5053876.

[106] Hand TW, Vujkovic-Cvijin I, Ridaura VK, Belkaid Y. Linking the microbiota, chronic disease, and the immune system. Trends Endocrinol Metab 2016;27(12):831–43. Available from: https://doi.org/10.1016/j.tem.2016.08.003 Epub 2016/09/14. PubMed PMID: 27623245; PMCID: PMC5116263.

[107] Hold GL, Hansen R. Impact of the gastrointestinal microbiome in health and disease: co-evolution with the host immune system. Curr Top Microbiol Immunol 2019;421:303–18. Available from: https://doi.org/10.1007/978-3-030-15138-6_12 Epub 2019/05/28. PubMed PMID: 31123894.

[108] Indrio F, Martini S, Francavilla R, Corvaglia L, Cristofori F, Mastrolia SA, et al. Epigenetic matters: the link between early nutrition, microbiome, and long-term health development. Front Pediatr 2017;5:178. Available from: https://doi.org/10.3389/fped.2017.00178 Epub 2017/09/08. PubMed PMID: 28879172; PMCID: PMC5572264.

[109] Ling Z, Li Z, Liu X, Cheng Y, Luo Y, Tong X, et al. Altered fecal microbiota composition associated with food allergy in infants. Appl Environ Microbiol 2014;80(8):2546–54. Available from: https://doi.org/10.1128/aem.00003-14 Epub 2014/02/18. PubMed PMID: 24532064; PMCID: PMC3993190.

[110] Lynch SV, Boushey HA. The microbiome and development of allergic disease. Curr Opin Allergy Clin Immunol 2016;16(2):165–71. Available from: https://doi.org/10.1097/aci.0000000000000255 Epub 2016/02/18. PubMed PMID: 26885707; PMCID: PMC5378446.

[111] Stewart CJ, Ajami NJ, O'Brien JL, Hutchinson DS, Smith DP, Wong MC, et al. Temporal development of the gut microbiome in early childhood from the TEDDY study. Nature 2018;562(7728):583–8. Available from: https://doi.org/10.1038/s41586-018-0617-x Epub 2018/10/26. PubMed PMID: 30356187; PMCID: PMC6415775.

[112] Granato A, Chen Y, Wesemann DR. Primary immunoglobulin repertoire development: time and space matter. Curr Opin Immunol 2015;33:126–31. Available from: https://doi.org/10.1016/j.coi.2015.02.011 Epub 2015/03/24. PubMed PMID: 25797714; PMCID: PMC4414491.

[113] Wesemann DR. Microbes and B cell development. Adv Immunol 2015;125:155–78. Available from: https://doi.org/10.1016/bs.ai.2014.09.005 Epub 2015/01/17. PubMed PMID: 25591467.

[114] Nguyen QN, Himes JE, Martinez DR, Permar SR. The impact of the gut microbiota on humoral immunity to pathogens and vaccination in early infancy. PLoS Pathog 2016;12(12):e1005997. Available from: https://doi.org/10.1371/journal.ppat.1005997 Epub 2016/12/23. PubMed PMID: 28006021; PMCID: PMC5179050.

[115] Madan JC, Koestler DC, Stanton BA, Davidson L, Moulton LA, Housman ML, et al. Serial analysis of the gut and respiratory microbiome in cystic fibrosis in infancy: interaction between intestinal and respiratory tracts and impact of nutritional exposures. MBio 2012;3(4). Available from: https://doi.org/10.1128/mBio.00251-12 Epub 2012/08/23. PubMed PMID: 22911969; PMCID: PMC3428694.

[116] Martin R, Nauta AJ, Ben Amor K, Knippels LM, Knol J, Garssen J. Early life: gut microbiota and immune development in infancy. Benef Microbes 2010;1(4):367–82. Available from: https://doi.org/10.3920/bm2010.0027 Epub 2011/08/13. PubMed PMID: 21831776.

[117] Backhed F. 99th Dahlem conference on infection, inflammation and chronic inflammatory disorders: the normal gut microbiota in health and disease. Clin Exp Immunol 2010;160(1):80–4. Available from: https://doi.org/10.1111/j.1365-2249.2010.04123.x Epub 2010/04/27. doi: CEI4123 [pii]. PubMed PMID: 20415855; PMCID: 2841839.

[118] Walker WA. The importance of appropriate initial bacterial colonization of the intestine in newborn, child, and adult health. Pediatric Res 2017;82(3):387–95. Available from: https://doi.org/10.1038/pr.2017.111 Epub 2017/04/21. PubMed PMID: 28426649; PMCID: PMC5570628.

[119] Ahern PP, Faith JJ, Gordon JI. Mining the human gut microbiota for effector strains that shape the immune system. Immunity 2014;40(6):815–23. Available from: https://doi.org/10.1016/j.immuni.2014.05.012 Epub 2014/06/21. PubMed PMID: 24950201; PMCID: PMC4118768.

[120] Walker WA. Bacterial colonization of the newborn gut, immune development, and prevention of disease. Nestle Nutr Inst Workshop Ser 2017;88:23–33. Available from: https://doi.org/10.1159/000455210 Epub 2017/03/28. PubMed PMID: 28346921.

[121] Ciabattini A, Olivieri R, Lazzeri E, Medaglini D. Role of the microbiota in the modulation of vaccine immune responses. Front Microbiol 2019;10:1305. Available from: https://doi.org/10.3389/fmicb.2019.01305 PubMed PMID: 31333592; PMCID: PMC6616116.

[122] Harris VC. The significance of the intestinal microbiome for vaccinology: from correlations to therapeutic applications. Drugs 2018;78(11):1063–72. Available from: https://doi.org/10.1007/s40265-018-0941-3 PubMed PMID: 29943376; PMCID: PMC6061423.

[123] Harris VC, Haak BW, Handley SA, Jiang B, Velasquez DE, Hykes Jr. BL, et al. Effect of antibiotic-mediated microbiome modulation on rotavirus vaccine immunogenicity: a human, randomized-control proof-of-concept trial. Cell Host Microbe 2018;24(2):197–207. Available from: https://doi.org/10.1016/j.chom.2018.07.005 e4 PubMed PMID: 30092197.

[124] Huda MN, Ahmad SM, Alam MJ, Khanam A, Kalanetra KM, Taft DH, et al. Bifidobacterium abundance in early infancy and vaccine response at 2 years of age. Pediatrics 2019;143(2). Available from: https://doi.org/10.1542/peds.2018-1489 PubMed PMID: 30674610; PMCID: PMC6361348 (California, Davis) has received grant support from Evolve Biosystems to fund a clinical trial of a probiotic in term infants; he has consulted for Avexegen and received payment for travel and lectures from Abbott. Dr Mills is a cofounder of and consultant for Evolve Biosystems and has stock and stock options therein; he has received payment for lectures from Nestle and Abbott; the other authors have indicated they have no potential conflicts of interest to disclose.

[125] Oh JZ, Ravindran R, Chassaing B, Carvalho FA, Maddur MS, Bower M, et al. TLR5-mediated sensing of gut microbiota is necessary for antibody responses to seasonal influenza vaccination. Immunity 2014;41(3):478–92. Available from: https://doi.org/10.1016/j.immuni.2014.08.009 PubMed PMID: 25220212; PMCID: PMC4169736.

[126] Zimmermann P, Curtis N. Factors that influence the immune response to vaccination. Clin Microbiol Rev 2019;32(2). Available from: https://doi.org/10.1128/CMR.00084-18 PubMed PMID: 30867162; PMCID: PMC6431125.

[127] Karst SM. The influence of commensal bacteria on infection with enteric viruses. Nat Rev Microbiol 2016;14(4):197–204. Available from: https://doi.org/10.1038/nrmicro.2015.25 Epub 2016/02/09. PubMed PMID: 26853118; PMCID: PMC5198578.

[128] Haas A, Zimmermann K, Graw F, Slack E, Rusert P, Ledergerber B, et al. Systemic antibody responses to gut commensal bacteria during chronic HIV-1 infection. Gut 2011;60 (11):1506–19. Available from: https://doi.org/10.1136/gut.2010.224774 Epub 2011/04/26. PubMed PMID: 21515549.

[129] Ngo ST, Steyn FJ, McCombe PA. Gender differences in autoimmune disease. Front Neuroendocrinol 2014;35(3):347–69. Available from: https://doi.org/10.1016/j.yfrne.2014.04.004 Epub 2014/05/06. PubMed PMID: 24793874.

[130] Klein SL, Flanagan KL. Sex differences in immune responses. Nat Rev Immunol 2016;16(10):626–38. Available from: https://doi.org/10.1038/nri.2016.90 Epub 2016/08/23. PubMed PMID: 27546235.

[131] Alexander KL, Targan SR, Elson 3rd CO. Microbiota activation and regulation of innate and adaptive immunity. Immunol Rev 2014;260(1):206–20. Available from: https://doi.org/10.1111/imr.12180 Epub 2014/06/20. PubMed PMID: 24942691; PMCID: PMC4080089.

[132] Bain CC, Cerovic V. Interactions of the microbiota with the mucosal immune system. Clin Exp Immunol 2020;199(1):9–11. Available from: https://doi.org/10.1111/cei.13400 Epub 2019/11/30. PubMed PMID: 31777062; PMCID: PMC6904583.

[133] Mueller S, Saunier K, Hanisch C, Norin E, Alm L, Midtvedt T, et al. Differences in fecal microbiota in different European study populations in relation to age, gender, and country: a cross-sectional study. Appl Environ Microbiol 2006;72(2):1027–33. Available from: https://doi.org/10.1128/aem.72.2.1027-1033.2006 Epub 2006/02/08. PubMed PMID: 16461645; PMCID: PMC1392899.

[134] Dominianni C, Sinha R, Goedert JJ, Pei Z, Yang L, Hayes RB, et al. Sex, body mass index, and dietary fiber intake influence the human gut microbiome. PLoS One 2015;10(4):e0124599. Available from: https://doi.org/10.1371/journal.pone.0124599 Epub 2015/04/16. PubMed PMID: 25874569; PMCID: PMC4398427.

[135] Haro C, Rangel-Zuniga OA, Alcala-Diaz JF, Gomez-Delgado F, Perez-Martinez P, Delgado-Lista J, et al. Intestinal microbiota is influenced by gender and body mass index. PLoS One 2016;11(5):e0154090. Available from: https://doi.org/10.1371/journal.pone.0154090 Epub 2016/05/27. PubMed PMID: 27228093; PMCID: PMC4881937.

[136] Yurkovetskiy L, Burrows M, Khan AA, Graham L, Volchkov P, Becker L, et al. Gender bias in autoimmunity is influenced by microbiota. Immunity 2013;39(2):400–12. Available from: https://doi.org/10.1016/j.immuni.2013.08.013 Epub 2013/08/27. PubMed PMID: 23973225; PMCID: PMC3822899.

[137] Elderman M, van Beek A, Brandsma E, de Haan B, Savelkoul H, de Vos P, et al. Sex impacts Th1 cells, Tregs, and DCs in both intestinal and systemic immunity in a mouse strain and location-dependent manner. Biol Sex Differ 2016;7:21. Available from: https://doi.org/10.1186/s13293-016-0075-9 Epub 2016/04/07. PubMed PMID: 27051505; PMCID: PMC4820953.

[138] Shastri P, McCarville J, Kalmokoff M, Brooks SP, Green-Johnson JM. Sex differences in gut fermentation and immune parameters in rats fed an oligofructose-supplemented diet. Biol Sex Differ 2015;6:13. Available from: https://doi.org/10.1186/s13293-015-0031-0 Epub 2015/08/08. PubMed PMID: 26251695; PMCID: PMC4527341.

[139] Fransen F, van Beek AA, Borghuis T, Meijer B, Hugenholtz F, van der Gaast-de Jongh C, et al. The impact of gut microbiota on gender-specific differences in immunity. Front Immunol 2017;8:754. Available from: https://doi.org/10.3389/fimmu.2017.00754 Epub 2017/07/18. PubMed PMID: 28713378; PMCID: PMC5491612.

[140] Markle JG, Frank DN, Adeli K, von Bergen M, Danska JS. Microbiome manipulation modifies sex-specific risk for autoimmunity. Gut Microbes 2014;5(4):485–93. Available from: https://doi.org/10.4161/gmic.29795 Epub 2014/07/10. PubMed PMID: 25007153.

[141] Markle JG, Frank DN, Mortin-Toth S, Robertson CE, Feazel LM, Rolle-Kampczyk U, et al. Sex differences in the gut microbiome drive hormone-dependent regulation of autoimmunity. Science 2013;339(6123):1084–8. Available from: https://doi.org/10.1126/science.1233521 Epub 2013/01/19. PubMed PMID: 23328391.

[142] Bolnick DI, Snowberg LK, Hirsch PE, Lauber CL, Org E, Parks B, et al. Individual diet has sex-dependent effects on vertebrate gut microbiota. Nature Commun 2014;5:4500. Available from: https://doi.org/10.1038/ncomms5500 Epub 2014/07/30. PubMed PMID: 25072318; PMCID: PMC4279269.

[143] Org E, Mehrabian M, Parks BW, Shipkova P, Liu X, Drake TA, et al. Sex differences and hormonal effects on gut microbiota composition in mice. Gut Microbes 2016;7 (4):313–22. Available from: https://doi.org/10.1080/19490976.2016.1203502 Epub 2016/06/30. PubMed PMID: 27355107; PMCID: PMC4988450.

Chapter 4

The microbiome in preterm infants and implications in health

Annie Z. Wang[1] and Erika C. Claud[2]
[1]*Department of Pediatrics, Section of Neonatology, The University of Chicago, Chicago, IL, United States,* [2]*Departments of Pediatrics, and Medicine, Sections of Neonatology and Gastroenterology, The University of Chicago, Chicago, IL, United States*

Introduction

Prematurity is a substantial worldwide health concern. The global rate of preterm birth is 10.6% of all births, and complications of preterm birth are the leading cause of death in children less than 5 years old [1]. It must be emphasized that the preterm infant is not simply a small baby. Instead, it is more accurate to view them as a fetus that must adapt early to the ex utero world, requiring prolonged medical support and hospitalization. All of their body systems are developmentally immature and forced to continue developing in a hostile environment. Thus preterm infants are at risk for a unique profile of diseases and complications, such as respiratory distress syndrome, bronchopulmonary dysplasia, intraventricular hemorrhage, necrotizing enterocolitis (NEC), and retinopathy of prematurity. Notably, many of these morbidities are at least partly inflammatory in nature, in keeping with the fact that preterm infants exist in a pro-inflammatory state. The immature immune system has decreased ability to protect the infant from pathogens, but is also unable to adequately downregulate the inflammatory response [2].

Understanding the microbiome specific to the preterm infant can lead to powerful tools in caring for this vulnerable population. These infants are at risk of dysbiosis for many reasons, including immaturity of the immune system, frequent exposure to broad-spectrum antibiotics, delays in enteral feeding, and exposures such as indwelling catheters and fomites encountered during prolonged hospitalization. Investigation of the microbiome as it develops in the preterm infant and how it participates in host health and disease,

The Developing Microbiome. DOI: https://doi.org/10.1016/B978-0-12-820602-7.00004-0

such as in NEC, is a critically important area of exploration. An overarching mechanism of the microbiome in host pathophysiology at all stages of life revolves around modulating barrier function and systemic inflammation [3]. More in-depth understanding may potentially reveal pathways for intervention on the microbiome to promote health both during the initial hospitalization and throughout life.

Characteristics of the preterm microbiome

Establishment of the microbiome

There has been debate in the scientific community between the sterile womb doctrine versus the hypothesis of in utero colonization [4]. Whether bacteria are introduced before or after delivery, it is known that colonization expands rapidly within hours of an infant being born. Facultative anaerobes are the usual pioneer species. These then generate a favorable environment for the establishment of obligate anaerobes, which dominate the microbiome of term infants thereafter [5–7]. The host also contributes to this transition as enterocytes consume oxygen within the intestinal lumen, in turn depriving microbes of oxygen [8]. Important commensal organisms include *Bifidobacterium*, *Bacteroides*, and *Akkermansia* [9–12].

Overall, the preterm gut microbiome has low diversity and richness with high interindividual variability, compared to term infants. It is enriched in taxa known to include potential pathogens, such as Enterobacteriaceae, *Staphylococcus*, *Enterococcus*, and *Clostridium* [13–19]. Preterm infants tend to have increased abundance of gram-negative species, many being members of Gammaproteobacteria, rather than the gram-positive Firmicutes and Actinobacteria that are found to be core members of the term microbiome (i.e., *Bifidobacterium*, *Lactobacillus*, *Propionibacterium*, and so on) [20,21]. Metabolic differences have also been reported, such as decreased levels of short-chain fatty acids (SCFAs) [16] and decreased signals of genes related to glycan metabolism [22].

Maturation of the microbiome with age

Just as the microbiome in term infants changes predictably over time, there is substantial evidence that the microbiome of preterm infants progresses through specific stages correlating with gestational age, converging toward a term-like composition. This is in keeping with the idea that maturation of the gut and immune system is key in cross talk with a normal microbiome. The immune system, including protective and adaptive functions, is immature in preterm infants, leading to both increased susceptibility to infection by virulent organisms and exaggerated inflammatory responses to bacteria [2,23,24]. Intestinal motility is decreased due to incomplete innervation of motility complexes. Barrier function, including tight junctions, secretory immunoglobulin A, and mucus production, are dysregulated in the preterm gut. Further, toll-like receptors (TLRs) on the gut epithelium that interact with bacteria are developmentally regulated [23,25]. These factors impact the way the immature host and microbiota interact and how the gastrointestinal tract is colonized. Importantly, in addition to maturation of the host, metabolic interactions between bacterial species may also contribute to the ordered succession of the microbial community.

The earliest stage of the microbiome is the least diverse. It is dominated by Firmicutes, particularly *Enterococcus*, *Streptococcus*, and *Staphylococcus* species. This is followed by an intermediate stage dominated by Gammaproteobacteria, primarily members of the Enterobacteriaceae family, which is then succeeded by the most mature stage with abundant Clostridia and high diversity [20,26–28]. Achievement of this most mature stage is delayed in preterm infants compared to full-term infants [27].

Bifidobacterium is normally dominant in term, breastfed infants. However, in a cohort of preterm infants, colonization with *Bifidobacterium* was only transient or present at low abundance in infants born at less than 32 weeks. *Bifidobacterium* colonization with high abundance was successful only in infants born after 32 weeks, and it was typically established around 34 weeks postconceptional age [29]. Other studies have also identified postnatal or postconceptional age as major determinants of microbiome composition in preterm infants [30–33]. These findings strengthen the case for the role of host maturation in the composition and progression of the fecal microbiome. In addition, the normal progression to an obligate anaerobe-dominated community is delayed [16,17,34,35], and infants born at earlier gestational ages have decreased abundance of obligate anaerobes [36]. In addition to host immaturity, investigators have suggested that exogenous oxygen administration, often required in preterm infants with underdeveloped lungs, may contribute to a more aerobic environment in the gut [37], promoting dysbiosis.

The composition of the microbial community becomes increasingly similar to that of a healthy term infant with increasing age [15,38]. A long-term follow-up study of former preterm infants found that at 1–3 years of age, common neonatal intensive care unit (NICU)-associated organisms such as *Klebsiella oxytoca* and *Enterococcus faecalis* had significantly decreased, and more mature organisms such as *Akkermansia*, *Blautia*, *Bacteroides*, and *Collinsella* had increased, with diversity at a level similar to that found in former term infants [39]. Similarly, at follow-up at age 4 years in former very low-birthweight (VLBW) infants, the child's fecal microbiota closely resembled their mothers' microbiota [40]. Thus although preterm infants have a distinct microbial composition in infancy, during childhood the community appears to recover, supporting the importance of host maturation in determining the makeup of the microbiome.

Environmental determinants of the preterm microbiome

Mode of delivery

Environmental factors are important in shaping the infant gut microbiome. Mode of delivery is known to have significant impact in term infants, with vaginal delivery promoting normal microbiome development. Cesarean delivery leads to decreased vertical transmission of maternal microbiota compared to vaginal birth. As a result, gut colonization after cesarean delivery occurs with skin and environmental microbes rather than maternal vaginal and gut microbes [41,42]. This is an important consideration in preterm infants, where cesarean delivery occurs more frequently [43]. However, the data have been heterogenous in preterm infants. Some studies have found no association between gut microbiome composition and mode of delivery [16,27,34,44]. La Rosa et al. showed some small effects of cesarean delivery on the pace of the ordered succession of the microbiome, but the overall trajectory was not affected [28]. Higher abundance of *Bifidobacterium* at 10 days of age in vaginally delivered preterm infants was identified in one study [45]. Stewart et al. also found that there was a divergence between the microbiota of cesarean and vaginally delivered infants that disappeared by 5 weeks of age; however, infants who were delivered vaginally demonstrated greater stability of their microbiota at 2 months and at follow-up after discharge from the hospital [46]. Thus mode of delivery appears to contribute minimally to the makeup of the gut microbiome in preterm infants. Developmental immaturity and frequent exposure to other environmental factors may overshadow the impact that cesarean delivery would have on the preterm microbiome.

Feeding: breast milk, formula, and enteral deprivation

Diet is another important modifier of the microbiome that is well-studied in full-term infants, focusing primarily on the effect of formula feeding versus breast milk feeding, as well as the introduction of solid foods. Colostrum and breast milk may contain communities of bacteria that are important probiotic organisms for the growing infant [47–51]. Substances in breast milk such as human milk oligosaccharides (HMOs) also act as important prebiotics, supporting the growth of commensal species. Thus breast milk feeding promotes eubiosis in the infant microbiome.

Maternal breast milk (MBM) is a vehicle for bacterial colonization in preterm infants as well. This was demonstrated by Stewart et al., who found that the microbiome obtained from infants' fecal samples shared a core community with their mothers' MBM, which is primarily composed of members of *Bacilli* and Gammaproteobacteria [52]. Breast milk-fed infants born at ≤ 32 weeks gestation had greater initial diversity and a more gradual increase in diversity over time compared to formula-fed infants. It is proposed that this pattern might be health-promoting as early diversity is widely thought to be beneficial, but rapid increases in the microbial population may be associated with increased inflammatory response. This group also found that MBM also attenuates the effects of gut immaturity on microbial colonization, as the microbiome in formula-fed infants clusters by birthweight, while no such association with birthweight is seen in breast milk-fed infants. The microbiome remains in immature stages longer in formula-fed infants than in their breast milk-fed counterparts [26].

Investigators also examined the effects of donor breast milk (DBM) on the microbiome. This is commonly given to very preterm infants in lieu of formula, in light of the large body of evidence for the protective effects of breast milk. The benefits are not the same as MBM, since DBM is pasteurized, pooled breast milk typically donated by mothers of older, term infants, which is nutritionally different from MBM of mothers of preterm infants. In this study, there was a differential effect based on gestational age at birth, in keeping with the importance of gut maturation. In infants born after 28 weeks, DBM promoted a microbiome closely resembling the microbiome of an MBM-fed infant. However, in infants born earlier than 28 weeks gestation, DBM did not achieve an MBM-like microbiome, but it promoted colonization by specific breast milk-associated taxa, such as Enterobacteriales [26]. A recent study of VLBW infants found that in comparison to those receiving MBM, infants receiving mostly DBM had decreased diversity and abundance of *Bifidobacterium* and *Bacteroides*. They had a greater abundance of *Staphylococcus* in the fecal microbiome [53].

Preterm infants are more likely than full-term infants to have periods of deprivation of enteral feeds. This is both due to cautious initiation of enteral feeds after birth and intermittent discontinuation of feeding with clinical illness or feeding intolerance during their hospital course. In the gut microbiome of an infant born at 28 weeks gestation, species richness and diversity decreased rapidly after feeds were discontinued due to feeding intolerance. During this time, the community was dominated by *Pseudomonas*, *Staphylococcus*, and other members of Pseudomonadaceae. After reinitiating feeds with formula, diversity and richness recovered in about 2 days, and the microbiome composition had shifted to predominant Enterobacteriaceae and *Enterococcus* [18]. Infants who were dependent on total parenteral nutrition were found to have lower diversity and lower abundance of Bacteroidetes in comparison to enterally fed controls. In this particular study, the median gestational age of subjects was 36 weeks, and infants receiving parenteral nutrition were more likely to have gastrointestinal pathology, undergo surgery, and receive antibiotics [54]. Our ability to generalize these results to a preterm population is limited, but it does underscore the finding that enteral deprivation alters gut microbial colonization in newborns.

Antibiotics: pre- and postnatal

The microbiome is known to change with antibiotics administration in term infants. Intrapartum antibiotic prophylaxis (IAP) of laboring mothers for Group B *Streptococcus* colonization is a common peripartum exposure to antibiotics. In both term and preterm infants, IAP reduces microbial richness and diversity and decreases the abundance of commensal organisms. Interestingly, IAP had an equal or greater effect on the microbiome when compared to direct administration of antibiotics to the infant [9,45]. Investigators also found differences in the activity of various functional pathways of the microbiome in preterm infants whose mothers received IAP, including cell growth and death, metabolism and biosynthesis, and replication and repair [22].

Broad-spectrum antibiotics are the most commonly prescribed medications in the NICU [55]. Their use is nearly universal in preterm infants due to their increased risk of infectious complications [56,57]. Greater duration of antibiotics in preterm infants in the absence of bacteremia is associated with increased risk of adverse outcomes such as NEC and mortality [58,59]. Thus the impact of antibiotics on the preterm gut microbiome is a particularly relevant area of study. Administration of antibiotics to preterm infants is associated with decreased microbial diversity and richness [32,52,60–62]. Antibiotics also delay maturation of the microbiome [27,36], with higher abundance of Proteobacteria, which is usually more prevalent at early stages, and decreased mature organisms from phylum Firmicutes, especially *Clostridium*, Actinobacteria, and Bacteroidetes [63]. Other studies have found that antibiotics decrease the abundance of Enterobacteriaceae, as well as *Enterococcus* [52,61]. Interestingly, when comparing the effects of a brief (1–4 days) and prolonged (5–7 days) course of antibiotics to nonexposed controls amongst preterm infants, the effect appeared to be dose dependent. The composition of the microbiome exposed to a brief course was intermediate between the prolonged course and controls. Furthermore, 3 weeks later, infants who had received a brief course of antibiotics had recovered, with a gut microbiome that closely resembled controls, while those who received a prolonged course remained divergent from controls [61].

Long-term follow-up of preterm infants showed that microbial maturity, which was initially delayed in the NICU after administration of antibiotics, had recovered by 12–15 months of age to resemble nonantibiotics-exposed infants [32]. Studies have demonstrated that antibiotic resistance increases in the microbiome when an infant receives antibiotics; at follow-up at 2 years of age, antibiotic-resistant strains had been replaced with susceptible community-derived strains [21]. Thus it appears that while antibiotics given for prolonged duration in preterm infants do modify the microbiome in the short term by decreasing diversity, delaying maturation, and increasing antibiotic resistance, the microbial community does eventually recover. Given the evidence in former term infants showing associations between antibiotics in infancy and childhood outcomes such as atopy and obesity [64–66], early changes in the microbiome may affect the health of the host organism in lasting ways despite normalization of the microbiota.

Acid suppression

Acid suppression medications—that is, histamine-2 receptor antagonists and proton pump inhibitors—have also been implicated in shaping the gut microbiome in preterm infants. These may be prescribed to preterm infants to aid with symptomatic reflux and feeding intolerance. Since these medications raise the pH of the gastrointestinal tract, which is an important facet of innate immunity, it is logical that they would change the microbiome. Evidence has shown greater risk of infection and NEC with the use of histamine-2 receptor blockers in infants in the NICU [67], but data on their impact on the microbiome in pediatrics is limited. There is ample adult data demonstrating unfavorable changes in the microbiome and increased enteric infections with the use of proton pump inhibitors [68]. *Clostridium difficile* infection in infants is positively associated with current or recent use of acid suppression medications [69], and in infants with GERD, proton pump inhibitors transiently altered the microbiome [70]. Some studies in preterm infants have included the use of acid suppression medications as a clinical variable: for instance, Grier found that these medications were associated with increased abundance of Proteobacteria, Actinobacteria, and Bacteroidetes [27]. To our knowledge, only one study targeting the impact of acid suppression medications on the preterm fecal microbiome has been published: Gupta et al. found that preterm infants receiving histamine-2 receptor antagonists had decreased diversity of the microbiome, decreased Firmicutes, and greater abundance of Proteobacteria, driven by increases in Enterobacteriaceae [71]. Further study is warranted to investigate how the microbiome participates in the association between acid suppression medications and adverse outcomes.

Hospitalization

The hospital environment cannot be discounted as a factor in the microbial colonization of the preterm infant. In term infants, even brief hospitalization is associated with higher abundance of *C. difficile* [9]. Preterm infants remain in the NICU for a significant period of time, sometimes months, and are exposed to many possible reservoirs of bacteria, despite thorough hygiene practices in hospitals [72]. Brooks et al. found bacteria residing on multiple sites in the NICU, including nurse station electronics, sinks, general work surfaces, feeding and intubation tubing, healthcare providers' and parents' hands, and access knobs on incubators. Feeding and intubation tubing had the highest relative abundance of microbiota normally found in preterm infant stool. One infant in this study appeared to reacquire the same strain of *Klebsiella pneumoniae* from the environment at multiple time points, after which it would be outcompeted in the gut, lending support to the possibility of colonization of the gut by environmental microbiota. The likelihood of patient-to-patient transmission via the environment was also demonstrated when a strain of *E. faecalis* was found to be shared between the study infants [14]. Similarly, in another single-unit study of infants who did and did not develop bloodstream infections, the identified pathogens were present in the fecal microbiota of both infected infants and other healthy controls residing in the NICU at the same time [73].

The role of the preterm gut microbiome in host health

Necrotizing enterocolitis

Neonatal NEC is an inflammatory disease of the premature infant characterized by ischemic necrosis of the intestine. It is one of the most devastating diseases in preterm infants, occurring in about 7% of VLBW infants with an estimated mortality of 20%–30% [74]. Age of onset is inversely proportional to gestational age at birth [75]. It may initially manifest with nonspecific symptoms such as feeding intolerance and abdominal distension, which can progress to overwhelming sepsis, peritonitis, bowel necrosis and perforation, and death. Survivors are at increased risk of microcephaly and neurodevelopmental delays. Treatment is primarily supportive: bowel rest and abdominal decompression, systemic antibiotics, and surgical interventions as necessary such as peritoneal drainage or bowel resection [74].

In order to discuss the role of the gut microbiome in NEC, it is necessary to briefly review the current understanding of the pathogenesis of NEC. Gaps in knowledge remain, but it is likely multifactorial, including compromised perfusion of the gut, immune system immaturity and dysregulation, gut immaturity, and dysbiosis. These promote intestinal injury, bacterial translocation, and systemic inflammation, as summarized in Fig. 4.1. First, perfusion via intestinal microcirculation is impaired in NEC, mediated by activation of endothelial TLR4, which occurs via translocation of bacterial products across the gut barrier. This signaling reduces expression of endothelial nitric oxide synthase, which normally promotes vasodilation. Compromised circulation leads to intestinal ischemia and necrosis, which are key components of NEC [76].

Secondly, an inappropriately robust inflammatory response is a vital contributor to NEC in premature infants [77]. A recurring finding is that TLR4 signaling, which is activated by bacterial lipopolysaccharide (LPS), promotes cellular responses that predispose toward NEC [36]. Intestinal immune cells are imbalanced toward a pro-inflammatory state, characterized by low levels

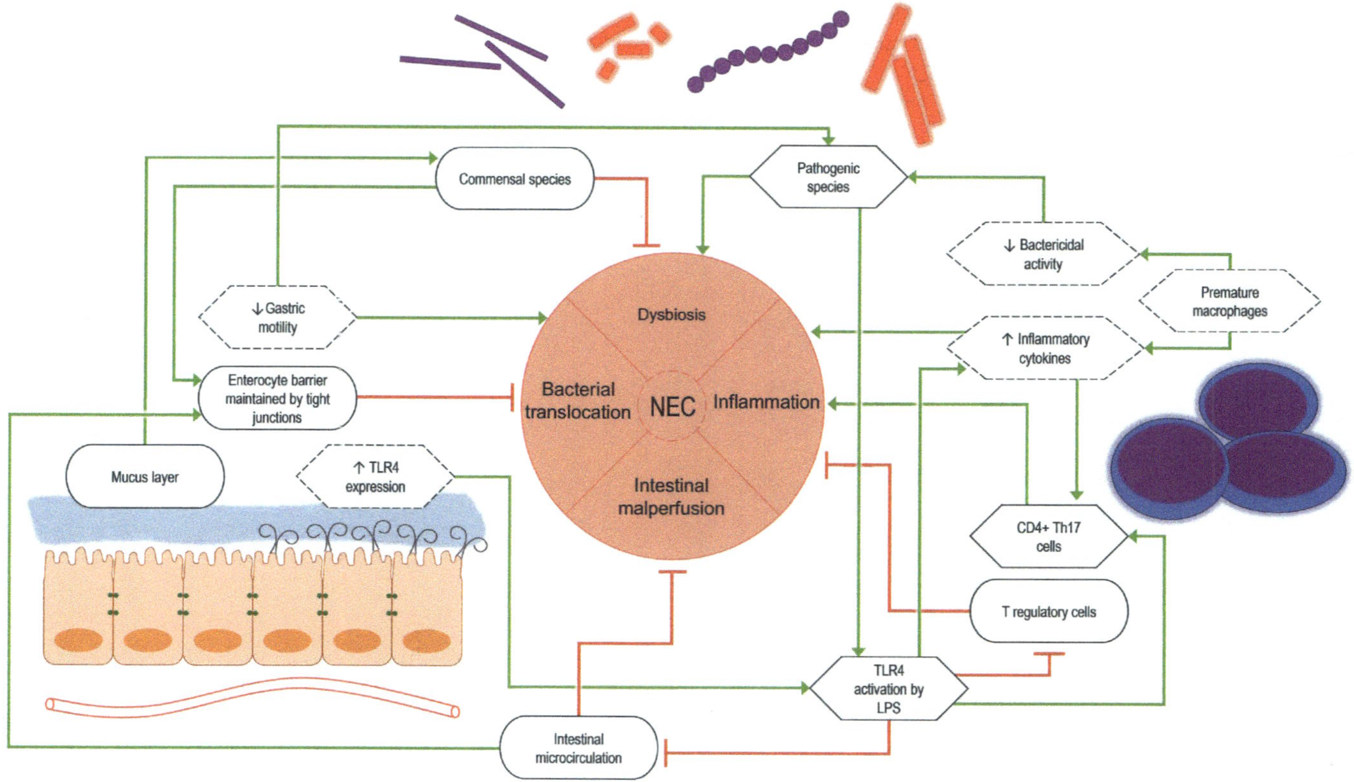

FIGURE 4.1 Necrotizing enterocolitis (NEC) occurs as a result of the interplay of dysbiosis, bacterial translocation, inflammation, and intestinal malperfusion. Multiple facets of the gastrointestinal tract, the immune system, and the microbiome in the preterm infant interact in a complex manner to influence this interplay. This diagram illustrates our current understanding of how some key factors prevent or promote NEC in premature infants. This is a simplification of current knowledge; also important are the roles of other immune cells and molecules, other gastrointestinal tissues and cell types, specific interactions between microbiota and the host, and genetic predispositions. Note the multiple ways in which TLR4 interacts with various host factors, after activation by LPS on microbes in the gut, to dispose the host toward NEC. For example, TLR4 decreases T regulatory cell production, which are counterinflammatory, while increasing Th17 cell production, which are pro-inflammatory. (*Rounded boxes* indicate factors protective against NEC. *Pointed boxes* indicate factors disposing toward NEC. *Dashed boxes* indicate factors specific to premature infants. TLR4, toll-like receptor 4; LPS, lipopolysaccharide).

of T regulatory cells and high levels of CD4+ Th17 cells. This pattern is promoted by TLR4 signaling, and it is associated with NEC [78]. Mice that lacked functional T and B cells were protected from NEC [79], enforcing the finding that NEC is, in part, immune cell mediated. Macrophages of premature infants have a more pro-inflammatory profile than those of full-term infants, having decreased phagocytic and bactericidal activity with increased inflammatory cytokine production, which is again exaggerated in NEC. Other immune cells and structures also play a role in intestinal homeostasis and in maintaining the gut barrier, such as neutrophils and dendritic cells. A heightened inflammatory state is also maintained by increased production of pro-inflammatory cytokines in the premature infant gut [80]. TLR4 is expressed at higher levels on enterocytes in prematurity, due to its role in normal gut development [23]. In addition to previously discussed mechanisms, it predisposes to NEC by promoting inflammation through nuclear factor-κβ activation, disruption of the intestinal barrier, and allowing bacterial translocation with resultant systemic inflammatory response [78].

The premature gastrointestinal tract itself is a key player in the pathogenesis of NEC. Decreased intestinal motility prolongs the dwell time of potential pathogens [25]. Enterocytes, the most numerous cells, constitute a physical barrier and also participate in the immune response by expression of pattern-recognition receptors such as TLRs. Tight junctions between enterocytes are vital to maintaining intestinal barrier function. Enterocyte apoptosis, disruption of tight junctions, decreased proliferation of enterocytes from stem cells, and impaired migration of newly differentiated enterocytes to damaged mucosa contribute to NEC; all of these are promoted by TLR4 activation. Goblet cells generate the mucus layer of the intestine, preventing interaction between pathogens and the epithelium; it also provides a habitat for commensal populations. Goblet cell depletion or inhibition in animal models increases susceptibility to and severity of NEC. Numerous other cell types also participate in host immunity and have been shown to play a role in promoting or protecting against NEC [78].

Finally, NEC is also a microbe-mediated disease. Live bacteria are necessary to induce experimental NEC in animal models [81]. However, the evidence for bacteria as the cause of NEC has been inconsistent over decades of research, suggesting that there is no single causative pathogen [82]. Known potential pathogens are frequently detected in the fecal microbiota of preterm infants, and blood cultures of infants with NEC do sometimes isolate infectious organisms, but in a minority of cases. Raveh-Sadka et al. underscored this with the finding that in an outbreak of NEC in a single NICU, few bacterial strains were found in more than one individual, and no bacterial strains were shared by infants with NEC [83]. It is increasingly understood that dysbiosis and bacterial translocation play an equal part in the storm of malperfusion, exaggerated inflammation, and gut immaturity that lead to NEC [24]. As an example, deficiencies of certain antimicrobial peptides lead to decreased Bacteroidetes and increased Firmicutes, considered to be an unfavorable change [78]. Administration of *Bifidobacterium* appeared to preserve occludin and claudin-4 at tight junctions in the gut epithelium, thus improving barrier function [84]. LPS, expressed by gram-negative species, is a ligand for TLR4, the underlying signal for many pathways leading to NEC.

Most studies of the microbiome in NEC, although heterogenous, support that NEC is associated with dysbiosis. While examining stools collected immediately before diagnosis of NEC, the microbiome was found to have reduced richness and diversity, with significantly increased dominance of Proteobacteria compared to controls. Further, in NEC cases, Proteobacteria tended to be dominated by a single genus, whereas controls tended to have a more even distribution within this phylum. Interestingly, the control infants who were exceptions to this balanced distribution of Proteobacteria went on to develop NEC after the study period [13]. A possible explanation of this association with NEC is that a bloom of a single potential pathogen activates the inflammatory cascade that leads to disease.

Other studies have also focused on fecal samples obtained before diagnosis, in order to characterize the dysbiosis that promotes NEC. Some studies [85,86] have not identified a consistent change in the microbiome. In others, a reduction in diversity is typically found before diagnosis [52,87]. A systemic meta-analysis found an increase in abundance of Proteobacteria and decreased abundance of Firmicutes and Bacteroidetes in NEC infants compared to controls; this divergence occurred around 30 weeks postconceptional age [63]. This finding continues to be corroborated by later studies [62]. Mai et al. found that abundance of Proteobacteria in early life was lower in NEC infants than in controls. They proposed that insufficient colonization at the early stage does not allow for normal immune tolerance to develop, predisposing the host to an exaggerated immune response to the subsequent bloom of Proteobacteria [88]. Specific members of Proteobacteria that have been identified in these studies included *Escherichia* [86], *Klebsiella* [52,89], and *Citrobacter* [19]. However, significant increases in taxa belonging to Firmicutes have also been found to precede NEC, such as *Enterococcus* [19], *Bacillus*, *Streptococcus*, and *Staphylococcus* [87]. Decreasing abundance of anaerobes such as Negativicutes or persistence of high Gammaproteobacteria over time—in other words, failure of the microbiome to mature as expected—has also been described in NEC [36]. The microbial composition of healthy preterm infants clusters by postnatal age, and over time increasingly resembles the microbiome of a term infant, but infants who developed NEC diverge from this progression beginning 3 weeks before NEC diagnosis [38].

Some studies found evidence for different patterns of inappropriate colonization leading to early versus late NEC, occurring before and after 3 weeks of life, respectively. Late NEC was associated with increased abundance of Gammaproteobacteria, and greater abundance of Clostridia was associated with early NEC [30]. Evidence elsewhere has also linked *Clostridium perfringens*-associated NEC with earlier onset, more severe clinical course, and higher mortality [37]. Firmicutes was correlated with early NEC in another study [90], mostly traced to *Staphylococcus* and *Enterococcus*. As to how specific patterns of dysbiosis may predispose to NEC, greater abundance of members of Proteobacteria likely increases LPS-sensing by TLR4 (Fig. 4.1). Abundant gram-positive organisms from Firmicutes may induce an overwhelming inflammatory response due to increased burden of peptidoglycan, the major component of the bacterial cell wall [90].

Late-onset sepsis

Another important complication of prematurity is late-onset sepsis (LOS), which is defined as bloodstream infection after 72 hours of age in a neonate [73]. Early-onset sepsis, which occurs before 72 hours, can typically be traced to organisms acquired in utero or during delivery. Organisms leading to LOS are likely derived from other sources. The infant gut has been suspected as a reservoir for pathogens causing LOS. In a single-unit study, fecal samples of 7 out of 11 infants who developed bloodstream infections contained the causative organism—*Streptococcus agalactiae*, *Serratia marcescens*, or *Escherichia coli* [73]. Before onset of disease in LOS, the sepsis-causative organism may be identified in stool [87,91] and may predominate the fecal microbiome [92].

Beyond establishing the gut as a potential source for infectious organisms, other studies have attempted to identify dysbiosis predisposing to LOS, like in NEC research. Taft did not find differences in the fecal microbiota that could be generalized across study centers [91], but others have found evidence for decreased maturity of the microbiota before onset of LOS. Specifically, infants who went on to develop LOS showed a failure to progress toward anaerobe dominance [92] and lack of mature taxa such as *Clostridium*, *Veillonella*, and, interestingly, *Klebsiella* [93].

Preterm infants are at increased risk of sepsis as the immune system is underdeveloped in multiple facets, sharing many factors with susceptibility to NEC (Fig. 4.1). Barrier function is diminished: the stratum corneum of the skin is underdeveloped, and various cell types of the gut and respiratory mucosa are immature, leading to impaired clearance of pathogens and decreased secretion of protective proteins and mucus [25]. Preterm infants receive less immunoglobulin G from transplacental transfer [94], and leukocyte function is impaired, particularly chemotaxis and bactericidal activity [2]. In addition, the healthy microbiome promotes gut epithelial integrity [92]. Thus the immaturity of the microbiome contributes to LOS in preterm infants in two ways: (1) by failing to promote gut barrier function, and (2) by harboring potential pathogens.

Growth and obesity

The microbiome participates in nutrient metabolism and growth [3], and growth is an important marker of overall wellness in the preterm infant. Studies of malnutrition and poor growth in infants and children have found associated changes in the microbiota, and animal models have shown that poor- and normal-growth phenotypes are transmitted by the microbiome [95–97]. This is likely both through metabolic functions contributing to nutrition and through promotion of host health. The abundance of certain taxa at early time points—Enterobacteriaceae and *Streptococcus* at day 2 and *Bacteroides* at day 10—had a positive association with weight gain at 1 month of age in preterm infants, again demonstrating that early differences in the microbiota, even those that are transient, have a lasting effect. Investigators also found that overall, the abundance of *Staphylococcus* and *Enterococcus* was negatively associated with weight gain, while there was a positive association with *Weisella* [98]. Younge et al. found that disrupted maturation of the microbiome was associated with growth failure in extremely preterm infants; these infants had persistently low diversity, prolonged dominance of Enterobacteriaceae, and decreased obligate anaerobes [99]. Other investigators corroborated these findings, demonstrating delayed transition out of early phases of microbiome development [27] and lower microbial diversity [33] in preterm infants with poor weight gain. Conversely, another study of VLBW infants found that gains in length—a marker of lean mass growth—were negatively associated with microbial maturity [40]. Clearly our understanding of the role of the microbiome in growth requires further study. Deeper insights into this interaction may uncover microbiome-targeted interventions to promote nutrition and growth in preterm infants.

On the other hand, preterm infants and those of low birthweight are at increased risk for obesity and metabolic syndrome later in life, particularly those with either inappropriately poor or rapid catch-up weight gain [3,80]. The microbiome during infancy has been implicated in childhood obesity in multiple studies of former term infants [64,100–102]. For instance, lower *Bifidobacterium* and higher *Staphylococcus aureus* in infancy are associated with later overweight status, which could be interpreted as the mechanism by which the microbiome mediates the link between formula feeding and obesity [103]. Further research is needed to investigate the microbiome as a modifiable risk factor for later obesity in preterm infants.

Neurodevelopment and the microbiome–gut–brain axis in the preterm infant

The microbiome–gut–brain axis is a complex interaction by which the microbiome effects change on brain development and function. Later chapters will discuss this phenomenon in greater detail. In brief, gnotobiotic animal models have shown that the presence of a gut microbiome is necessary for the normal development of the brain [104]. Germ-free subjects demonstrate altered behaviors and perform worse in motor and learning tasks. Their brain structures are immature compared to specific pathogen-free animals (which are presumed to have a normal microbiota), and gene expression in various brain regions is altered [105–107]. Germ-free mice that were transfaunated with microbiota from a preterm infant with poor growth developed decreased expression of neuronal and myelination markers, as well as increased neuroinflammation and altered neurotransmission pathways. This was in comparison to germ-free mice that had received microbiota from a preterm infant with good growth [108]. This underscores the importance of the microbiome as a player in brain development and as a mediator for other factors. The exact pathways through which the microbiome exerts its effect are not yet fully understood, but neural signaling through the enteric nervous system and vagus nerve, metabolic by-products of bacteria, modification of blood–brain barrier permeability, and immune activation have all been implicated [109–111].

The importance of the microbiome in the neurodevelopment of preterm infants cannot be understated. Gestational age at birth is inversely proportional to risk of neurodevelopmental impairment [112], and about 35%–50% of very preterm infants remain affected at school age [113]. Even late preterm infants, born at 34–36 weeks gestation, are at increased risk of impairment compared to term infants [114]. Deficits can persist into adulthood, although studies have demonstrated that developmental trajectories can change, for better or for worse, particularly in light of environmental factors such as socioeconomic status [115]. Major disabilities, including moderate/severe intellectual disability, hearing loss or blindness, cerebral palsy, and epilepsy, are increasingly frequent at lower birthweights. Risk of death or major disability increases with clinical morbidities, particularly bronchopulmonary dysplasia, brain injury, and/or retinopathy of prematurity [116]. Moreover, low severity impairments are increasingly recognized in formerly preterm children, such as borderline to low IQ scores, attention deficit hyperactivity disorder-like behaviors, autistic spectrum symptoms, executive function deficits, and behavior problems, affecting as many as 50%–70% of former-VLBW infants [113,117]. The explanation behind this increased incidence of neurodevelopmental impairment in preterm infants is manifold. Preterm birth likely disrupts normal in utero brain development, including corticogenesis, myelination, and blood–brain barrier formation, which continue through late gestation [118]. Preterm infants are also vulnerable to neurologic injury during their hospital course in the setting of increased inflammation, multiple morbidities of prematurity, and risk of intraventricular hemorrhage and periventricular leukomalacia [117]. Due to these increased risks, understanding the role of the microbiome in neurodevelopment of the preterm infant is a valuable area of research.

Interventions

Probiotics

Given the vulnerability of preterm infants to dysbiosis and its significant impact on the infant's health, modification of the microbiome to a desirable composition or functional state is an area of significant interest. Besides minimizing practices that predispose to potentially harmful organisms, active intervention is also a viable approach. Probiotics are defined as "live microorganisms that, when administered in adequate amounts, confer a health benefit on the host" [119]. Probiotic organisms do successfully colonize preterm infants [120]. Even when probiotics are administered to pregnant mothers and stopped after delivery, the probiotic organism remains detectable at up to 24 months of age [121]. The overall composition of the microbiome is not significantly changed by colonization with these organisms, but the metabolite profile of the community is altered, which may be an important aspect of how probiotics exert their effects [122].

In preterm infants, the main application of probiotics has been to reduce NEC. Proposed mechanisms of action include increasing the barrier to translocation of bacterial and metabolic by-products into the circulation, outcompeting potential pathogens, modification of host immune response, and augmentation of the mucosal IgA response [123]. A systematic meta-analysis of enteral probiotics found a significant reduction in incidence of NEC and mortality. The included studies used probiotics consisting of *Lactobacillus* alone or in combination with *Bifidobacterium* [124]. A more recent meta-analysis specifically focused on LOS did report a significant reduction in risk of LOS in preterm infants receiving probiotics [125].

Evidence for the benefits of probiotics in preterm infants continues to grow, meriting continued investigation of this therapy. However, due to considerable heterogeneity in formulation and administration protocols in published trials, it is difficult to make recommendations for probiotic administration in the routine care of preterm infants. Questions remain, including how probiotics actually impact the preterm infant microbiome immediately and over time, and whether intentional introduction of a high abundance of microorganisms is truly desirable in an immature host [126,127]. Considering the microbiome's known effects on priming the immune system and brain development at early stages, it is unclear whether probiotics would have a helpful or harmful effect in the long term. Clinicians have also expressed concerns surrounding the regulation of probiotic products to ensure safety, particularly since premature infants have compromised immune systems. A probiotic contaminated with *Rhizopus oryzae* was determined to be the origin of fatal mucormycosis in a preterm infant in 2014 [128]. There are case reports of probiotics-associated sepsis in infants, although this complication has not been reported in published trials [129,130]. A randomized controlled trial in adults with acute pancreatitis showed greater risk of mortality and bowel ischemia with probiotics versus placebo. It was proposed that the suddenly increased oxygen demand following the introduction of a high abundance of flora in an already compromised gut could have led to bowel ischemia [131], which has interesting theoretical applications to preterm infants, in whom gut malperfusion predisposes to NEC.

Prebiotics

Administering live bacteria to a vulnerable gut raises safety concerns, even if those are not robustly reflected in existing evidence. Instead, some investigators are studying prebiotics as an alternative way to promote eubiosis in preterm infants. A prebiotic is "a substrate that is selectively utilized by host microorganisms conferring a health benefit" [132]. Major substances being studied are both common in breast milk: lactoferrin and HMOs. Lactoferrin is thought to modulate inflammation and promote growth of probiotic bacteria [123]. A systematic review of lactoferrin administration to preterm infants demonstrated a reduction of risk of NEC and LOS, without affecting mortality. This benefit was magnified when lactoferrin was given in conjunction with probiotics [133].

HMOs are nondigestible compounds found in human breast milk; gut microbiota such as *Bifidobacterium* ferment HMOs into metabolites of use to the host, such as SCFAs. A meta-analysis of prebiotic oligosaccharide supplementation to infants born before 37 weeks gestation found no effect on risk of NEC, LOS, or time to reach full enteral feeds, although the prebiotics promoted greater abundance of *Bifidobacterium* [134]. Preterm animal models have also shown limited evidence of benefit. Investigators have questioned whether prebiotics supplementation, specifically HMOs, is appropriate in preterm infants. HMO content of MBM from mothers of preterm infants is widely variable [135], so a one-dose-fits-all approach may not be effective in this population. Maturational stage of the gut should also be considered, as the immature microbiota in the preterm gut may not be capable of benefiting from HMO supplementation.

Implications for lifetime health

The preterm infant is forced to adapt to and mature in an environment for which it is developmentally unsuited. This leads not only to unique morbidities of prematurity such as NEC and bronchopulmonary dysplasia, but also to long-term changes that impact health throughout life. Neurodevelopmental impairment was discussed earlier, affecting a large fraction of surviving preterm infants, particularly those suffering from more morbidities in the hospital course. Multiple studies have demonstrated that infants born prematurely are at increased risk for metabolic syndrome, ischemic cardiovascular disease, and cerebrovascular disease in adulthood [80,113,136]. Preterm infants are also at increased risk of obstructive or hyperreactive respiratory disorders throughout life, demonstrating greater need for inhaled asthma medications and greater risk for hospitalization. This was even observed in former preterm infants who were not diagnosed with bronchopulmonary dysplasia, suggesting that respiratory maldevelopment results from preterm birth, not just postnatal insults [137].

It is known that changes in the microbiome in early life can have both immediate and long-lasting effects on host health. If we can modify the microbiome in favorable ways, can we improve outcomes for infants born prematurely? A healthy microbiome protects against systemic inflammation, which is a significant contributor to the wide array of problems preterm infants face, and promotes normal neurodevelopment and growth. Interventions for the microbiome could potentially reduce the incidence of adverse outcomes such as NEC and LOS. By reducing early morbidities, it may also be possible to protect these infants from adverse outcomes later in life. Further, many morbidities in later life—asthma and atopic disease, obesity, metabolic syndrome, inflammatory bowel disease, and autism spectrum disorder and other neuropsychiatric disorders—have been associated with inappropriate inflammation and dysbiosis. These disorders will be discussed in detail elsewhere in this text, but it is valuable to state here that studying the microbiome in diseases of the preterm infant may (1) provide insights into the role of inflammation and dysbiosis in disorders such as these; and (2) reveal interventional targets for early prevention of these disorders.

Given our current understanding of the preterm microbiome, there are simple steps providers can take to promote eubiosis. Breast milk feeding should be encouraged when feasible, even with DBM. Medications that may alter the microbiome such as antibiotics and acid suppression medications should be prescribed judiciously. Despite knowledge gaps and questions about safety, probiotics and prebiotics are promising interventions that may come into use in the future. It remains to be seen whether these promote the optimal microbiome for the premature gut, and whether a standardized formulation and regimen can be recommended. Other modes of intervention, such as dietary changes, and utilizing the microbiome as a marker of health may become possible with deeper understanding. With further investigation, modification of the microbiome can become a valuable part of caring for the premature infant, both in the NICU and throughout life.

References

[1] Chawanpaiboon S, Vogel JP, Moller A-B, Lumbiganon P, Petzold M, Hogan D, et al. Global, regional, and national estimates of levels of preterm birth in 2014: a systematic review and modelling analysis. Lancet Glob Health 2019;7:e37–46. Available from: https://doi.org/10.1016/S2214-109X(18)30451-0.

[2] Melville JM, Moss TJM. The immune consequences of preterm birth. Front Neurosci 2013;7. Available from: https://doi.org/10.3389/fnins.2013.00079.

[3] Groer MW, Gregory KE, Louis-Jacques A, Thibeau S, Walker WA. The very low birth weight infant microbiome and childhood health. Birth Defects Res C Embryo Today 2015;105:252–64. Available from: https://doi.org/10.1002/bdrc.21115.

[4] Perez-Muñoz ME, Arrieta M-C, Ramer-Tait AE, Walter J. A critical assessment of the "sterile womb" and "in utero colonization" hypotheses: implications for research on the pioneer infant microbiome. Microbiome 2017;5:48. Available from: https://doi.org/10.1186/s40168-017-0268-4.

[5] Sender R, Fuchs S, Milo R. Revised estimates for the number of human and bacteria cells in the body. PLoS Biol 2016;14:e1002533. Available from: https://doi.org/10.1371/journal.pbio.1002533.

[6] Hansen R, Scott KP, Khan S, Martin JC, Berry SH, Stevenson M, et al. First-pass meconium samples from healthy term vaginally-delivered neonates: an analysis of the microbiota. PLoS One 2015;10:e0133320. Available from: https://doi.org/10.1371/journal.pone.0133320.

[7] Jost T, Lacroix C, Braegger CP, Chassard C. New insights in gut microbiota establishment in healthy breast fed neonates. PLoS One 2012;7:e44595. Available from: https://doi.org/10.1371/journal.pone.0044595.

[8] Byndloss MX, Pernitzsch SR, Bäumler AJ. Healthy hosts rule within: ecological forces shaping the gut microbiota. Mucosal Immunol 2018;11:1299–305. Available from: https://doi.org/10.1038/s41385-018-0010-y.

[9] Penders J, Thijs C, Vink C, Stelma FF, Snijders B, Kummeling I, et al. Factors influencing the composition of the intestinal microbiota in early infancy. Pediatrics 2006;118:511–21. Available from: https://doi.org/10.1542/peds.2005-2824.

[10] Vaishampayan PA, Kuehl JV, Froula JL, Morgan JL, Ochman H, Francino MP. Comparative metagenomics and population dynamics of the gut microbiota in mother and infant. Genome Biol Evol 2010;2:53–66. Available from: https://doi.org/10.1093/gbe/evp057.

[11] Yassour M, Vatanen T, Siljander H, Hämäläinen A-M, Härkönen T, Ryhänen SJ, et al. Natural history of the infant gut microbiome and impact of antibiotic treatment on bacterial strain diversity and stability. Sci Transl Med 2016;8:343ra81. Available from: https://doi.org/10.1126/scitranslmed.aad0917.

[12] Jakobsson HE, Abrahamsson TR, Jenmalm MC, Harris K, Quince C, Jernberg C, et al. Decreased gut microbiota diversity, delayed Bacteroidetes colonisation and reduced Th1 responses in infants delivered by caesarean section. Gut 2014;63:559–66. Available from: https://doi.org/10.1136/gutjnl-2012-303249.

[13] Wang Y, Hoenig JD, Malin KJ, Qamar S, Petrof EO, Sun J, et al. 16S rRNA gene-based analysis of fecal microbiota from preterm infants with and without necrotizing enterocolitis. ISME J 2009;3:944–54. Available from: https://doi.org/10.1038/ismej.2009.37.

[14] Brooks B, Firek BA, Miller CS, Sharon I, Thomas BC, Baker R, et al. Microbes in the neonatal intensive care unit resemble those found in the gut of premature infants. Microbiome 2014;2:1. Available from: https://doi.org/10.1186/2049-2618-2-1.

[15] Costello EK, Carlisle EM, Bik EM, Morowitz MJ, Relman DA. Microbiome assembly across multiple body sites in low-birthweight infants. MBio 2013;4. Available from: https://doi.org/10.1128/mBio.00782-13 e00782-00713.

[16] Arboleya S, Binetti A, Salazar N, Fernández N, Solís G, Hernández-Barranco A, et al. Establishment and development of intestinal microbiota in preterm neonates. FEMS Microbiol Ecol 2012;79:763–72. Available from: https://doi.org/10.1111/j.1574-6941.2011.01261.x.

[17] Magne F, Abély M, Boyer F, Morville P, Pochart P, Suau A. Low species diversity and high interindividual variability in faeces of preterm infants as revealed by sequences of 16S rRNA genes and PCR-temporal temperature gradient gel electrophoresis profiles. FEMS Microbiol Ecol 2006;57:128–38. Available from: https://doi.org/10.1111/j.1574-6941.2006.00097.x.

[18] Morowitz MJ, Denef VJ, Costello EK, Thomas BC, Poroyko V, Relman DA, et al. Strain-resolved community genomic analysis of gut microbial colonization in a premature infant. Proc Natl Acad Sci U S A 2011;108:1128–33. Available from: https://doi.org/10.1073/pnas.1010992108.

[19] Mshvildadze M, Neu J, Shuster J, Theriaque D, Li N, Mai V. Intestinal microbial ecology in premature infants assessed with non-culture-based techniques. J Pediatr 2010;156:20–5. Available from: https://doi.org/10.1016/j.jpeds.2009.06.063.

[20] Ho TTB, Groer MW, Kane B, Yee AL, Torres BA, Gilbert JA, et al. Dichotomous development of the gut microbiome in preterm infants. Microbiome 2018;6:157. Available from: https://doi.org/10.1186/s40168-018-0547-8.

[21] Moles L, Gómez M, Heilig H, Bustos G, Fuentes S, de Vos W, et al. Bacterial diversity in meconium of preterm neonates and evolution of their fecal microbiota during the first month of life. PLoS One 2013;8:e66986. Available from: https://doi.org/10.1371/journal.pone.0066986.

[22] Arboleya S, Sánchez B, Solís G, Fernández N, Suárez M, Hernández-Barranco AM, et al. Impact of prematurity and perinatal antibiotics on the developing intestinal microbiota: a functional inference study. Int J Mol Sci 2016;17. Available from: https://doi.org/10.3390/ijms17050649.

[23] Caicedo RA, Schanler RJ, Li N, Neu J. The developing intestinal ecosystem: implications for the neonate. Pediatr Res 2005;58:625–8. Available from: https://doi.org/10.1203/01.PDR.0000180533.09295.84.

[24] Claud EC, Walker WA. Hypothesis: inappropriate colonization of the premature intestine can cause neonatal necrotizing enterocolitis. FASEB J 2001;15:1398–403. Available from: https://doi.org/10.1096/fj.00-0833hyp.

[25] Collins A, Weitkamp J-H, Wynn JL. Why are preterm newborns at increased risk of infection? Arch Dis Child Fetal Neonatal Ed 2018;103:F391–4. Available from: https://doi.org/10.1136/archdischild-2017-313595.

[26] Gregory KE, Samuel BS, Houghteling P, Shan G, Ausubel FM, Sadreyev RI, et al. Influence of maternal breast milk ingestion on acquisition of the intestinal microbiome in preterm infants. Microbiome 2016;4:68. Available from: https://doi.org/10.1186/s40168-016-0214-x.

[27] Grier A, Qiu X, Bandyopadhyay S, Holden-Wiltse J, Kessler HA, Gill AL, et al. Impact of prematurity and nutrition on the developing gut microbiome and preterm infant growth. Microbiome 2017;5:158. Available from: https://doi.org/10.1186/s40168-017-0377-0.

[28] La Rosa PS, Warner BB, Zhou Y, Weinstock GM, Sodergren E, Hall-Moore CM, et al. Patterned progression of bacterial populations in the premature infant gut. Proc Natl Acad Sci U S A 2014;111:12522–7. Available from: https://doi.org/10.1073/pnas.1409497111.

[29] Butel M-J, Suau A, Campeotto F, Magne F, Aires J, Ferraris L, et al. Conditions of bifidobacterial colonization in preterm infants: a prospective analysis. J Pediatr Gastroenterol Nutr 2007;44:577–82. Available from: https://doi.org/10.1097/MPG.0b013e3180406b20.

[30] Zhou Y, Shan G, Sodergren E, Weinstock G, Walker WA, Gregory KE. Longitudinal analysis of the premature infant intestinal microbiome prior to necrotizing enterocolitis: a case-control study. PLoS One 2015;10:e0118632. Available from: https://doi.org/10.1371/journal.pone.0118632.

[31] Chernikova DA, Madan JC, Housman ML, Zain-Ul-Abideen M, Lundgren SN, Morrison HG, et al. The premature infant gut microbiome during the first 6 weeks of life differs based on gestational maturity at birth. Pediatr Res 2018;84:71–9. Available from: https://doi.org/10.1038/s41390-018-0022-z.

[32] Gasparrini AJ, Crofts TS, Gibson MK, Tarr PI, Warner BB, Dantas G. Antibiotic perturbation of the preterm infant gut microbiome and resistome. Gut Microbes 2016;7:443–9. Available from: https://doi.org/10.1080/19490976.2016.1218584.

[33] Jacquot A, Neveu D, Aujoulat F, Mercier G, Marchandin H, Jumas-Bilak E, et al. Dynamics and clinical evolution of bacterial gut microflora in extremely premature patients. J Pediatr 2011;158:390–6. Available from: https://doi.org/10.1016/j.jpeds.2010.09.007.

[34] Patel AL, Mutlu EA, Sun Y, Koenig L, Green S, Jakubowicz A, et al. Longitudinal survey of microbiota in hospitalized preterm very-low-birth-weight infants. J Pediatr Gastroenterol Nutr 2016;62:292–303. Available from: https://doi.org/10.1097/MPG.0000000000000913.

[35] Arboleya S, Solís G, Fernández N, de los Reyes-Gavilán CG, Gueimonde M. Facultative to strict anaerobes ratio in the preterm infant microbiota. Gut Microbes 2012;3:583–8. Available from: https://doi.org/10.4161/gmic.21942.

[36] Warner BB, Deych E, Zhou Y, Hall-Moore C, Weinstock GM, Sodergren E, et al. Gut bacteria dysbiosis and necrotising enterocolitis in very low birthweight infants: a prospective case-control study. Lancet 2016;387:1928–36. Available from: https://doi.org/10.1016/S0140-6736(16)00081-7.

[37] Sim K, Shaw AG, Randell P, Cox MJ, McClure ZE, Li M-S, et al. Dysbiosis anticipating necrotizing enterocolitis in very premature infants. Clin Infect Dis 2015;60:389–97. Available from: https://doi.org/10.1093/cid/ciu822.

[38] Claud EC, Keegan KP, Brulc JM, Lu L, Bartels D, Glass E, et al. Bacterial community structure and functional contributions to emergence of health or necrotizing enterocolitis in preterm infants. Microbiome 2013;1:20. Available from: https://doi.org/10.1186/2049-2618-1-20.

[39] Stewart CJ, Skeath T, Nelson A, Fernstad SJ, Marrs ECL, Perry JD, et al. Preterm gut microbiota and metabolome following discharge from intensive care. Sci Rep 2015;5:17141. Available from: https://doi.org/10.1038/srep17141.

[40] Yee AL, Miller E, Dishaw LJ, Gordon JM, Ji M, Dutra S, et al. Longitudinal microbiome composition and stability correlate with increased weight and length of very-low-birth-weight infants. mSystems 2019;4. Available from: https://doi.org/10.1128/mSystems.00229-18.

[41] Bäckhed F, Roswall J, Peng Y, Feng Q, Jia H, Kovatcheva-Datchary P, et al. Dynamics and stabilization of the human gut microbiome during the first year of life. Cell Host Microbe 2015;17:690–703. Available from: https://doi.org/10.1016/j.chom.2015.04.004.

[42] Dominguez-Bello MG, Costello EK, Contreras M, Magris M, Hidalgo G, Fierer N, et al. Delivery mode shapes the acquisition and structure of the initial microbiota across multiple body habitats in newborns. Proc Natl Acad Sci U S A 2010;107:11971–5. Available from: https://doi.org/10.1073/pnas.1002601107.

[43] Forsgren M, Isolauri E, Salminen S, Rautava S. Late preterm birth has direct and indirect effects on infant gut microbiota development during the first six months of life. Acta Paediatr 2017;106:1103–9. Available from: https://doi.org/10.1111/apa.13837.

[44] Hill CJ, Lynch DB, Murphy K, Ulaszewska M, Jeffery IB, O'Shea CA, et al. Evolution of gut microbiota composition from birth to 24 weeks in the INFANTMET Cohort. Microbiome 2017;5:4. Available from: https://doi.org/10.1186/s40168-016-0213-y.

[45] Arboleya S, Sánchez B, Milani C, Duranti S, Solís G, Fernández N, et al. Intestinal microbiota development in preterm neonates and effect of perinatal antibiotics. J Pediatr 2015;166:538–44. Available from: https://doi.org/10.1016/j.jpeds.2014.09.041.

[46] Stewart CJ, Embleton ND, Clements E, Luna PN, Smith DP, Fofanova TY, et al. Cesarean or vaginal birth does not impact the longitudinal development of the gut microbiome in a cohort of exclusively preterm infants. Front Microbiol 2017;8:1008. Available from: https://doi.org/10.3389/fmicb.2017.01008.

[47] Heikkilä MP, Saris PEJ. Inhibition of *Staphylococcus aureus* by the commensal bacteria of human milk. J Appl Microbiol 2003;95:471–8.

[48] Hunt KM, Foster JA, Forney LJ, Schütte UME, Beck DL, Abdo Z, et al. Characterization of the diversity and temporal stability of bacterial communities in human milk. PLoS One 2011;6:e21313. Available from: https://doi.org/10.1371/journal.pone.0021313.

[49] Jiménez E, Delgado S, Fernández L, García N, Albújar M, Gómez A, et al. Assessment of the bacterial diversity of human colostrum and screening of staphylococcal and enterococcal populations for potential virulence factors. Res Microbiol 2008;159:595–601. Available from: https://doi.org/10.1016/j.resmic.2008.09.001.

[50] Martín R, Jiménez E, Heilig H, Fernández L, Marín ML, Zoetendal EG, et al. Isolation of bifidobacteria from breast milk and assessment of the bifidobacterial population by PCR-denaturing gradient gel electrophoresis and quantitative real-time PCR. Appl Env Microbiol 2009;75:965–9. Available from: https://doi.org/10.1128/AEM.02063-08.

[51] Martín R, Olivares M, Marín ML, Fernández L, Xaus J, Rodríguez JM. Probiotic potential of 3 *Lactobacilli* strains isolated from breast milk. J Hum Lact 2005;21:8–17. Available from: https://doi.org/10.1177/0890334404272393 quiz 18–21, 41.

[52] Stewart CJ, Marrs ECL, Nelson A, Lanyon C, Perry JD, Embleton ND, et al. Development of the preterm gut microbiome in twins at risk of necrotising enterocolitis and sepsis. PLoS One 2013;8:e73465. Available from: https://doi.org/10.1371/journal.pone.0073465.

[53] Ford SL, Lohmann P, Preidis GA, Gordon PS, O'Donnell A, Hagan J, et al. Improved feeding tolerance and growth are linked to increased gut microbial community diversity in very-low-birth-weight infants fed mother's own milk compared with donor breast milk. Am J Clin Nutr 2019;109:1088–97. Available from: https://doi.org/10.1093/ajcn/nqz006.

[54] Dahlgren AF, Pan A, Lam V, Gouthro KC, Simpson PM, Salzman NH, et al. Longitudinal changes in the gut microbiome of infants on total parenteral nutrition. Pediatr Res 2019;86:107–14. Available from: https://doi.org/10.1038/s41390-019-0391-y.

[55] Hsieh EM, Hornik CP, Clark RH, Laughon MM, Benjamin DK, Smith PB, et al. Medication use in the neonatal intensive care unit. Am J Perinatol 2014;31:811–21. Available from: https://doi.org/10.1055/s-0033-1361933.

[56] Clark RH, Bloom BT, Spitzer AR, Gerstmann DR. Reported medication use in the neonatal intensive care unit: data from a large national data set. Pediatrics 2006;117:1979–87. Available from: https://doi.org/10.1542/peds.2005-1707.

[57] Flannery DD, Ross RK, Mukhopadhyay S, Tribble AC, Puopolo KM, Gerber JS. Temporal trends and center variation in early antibiotic use among premature infants. JAMA Netw Open 2018;1. Available from: https://doi.org/10.1001/jamanetworkopen.2018.0164.

[58] Cotten CM, Taylor S, Stoll B, Goldberg RN, Hansen NI, Sánchez PJ, et al. Prolonged duration of initial empirical antibiotic treatment is associated with increased rates of necrotizing enterocolitis and death for extremely low birth weight infants. Pediatrics 2009;123:58–66. Available from: https://doi.org/10.1542/peds.2007-3423.

[59] Kuppala VS, Meinzen-Derr J, Morrow AL, Schibler KR. Prolonged initial empirical antibiotic treatment is associated with adverse outcomes in premature infants. J Pediatr 2011;159:720–5. Available from: https://doi.org/10.1016/j.jpeds.2011.05.033.

[60] Gewolb IH, Schwalbe RS, Taciak VL, Harrison TS, Panigrahi P. Stool microflora in extremely low birthweight infants. Arch Dis Child Fetal Neonatal Ed 1999;80:F167–73. Available from: https://doi.org/10.1136/fn.80.3.f167.

[61] Greenwood C, Morrow AL, Lagomarcino AJ, Altaye M, Taft DH, Yu Z, et al. Early empiric antibiotic use in preterm infants is associated with lower bacterial diversity and higher relative abundance of *Enterobacter*. J Pediatr 2014;165:23–9. Available from: https://doi.org/10.1016/j.jpeds.2014.01.010.

[62] Lindberg TP, Caimano MJ, Hagadorn JI, Bennett EM, Maas K, Brownell EA, et al. Preterm infant gut microbial patterns related to the development of necrotizing enterocolitis. J Matern Fetal Neonatal Med 2020;33:349–58. Available from: https://doi.org/10.1080/14767058.2018.1490719.

[63] Pammi M, Cope J, Tarr PI, Warner BB, Morrow AL, Mai V, et al. Intestinal dysbiosis in preterm infants preceding necrotizing enterocolitis: a systematic review and meta-analysis. Microbiome 2017;5:31. Available from: https://doi.org/10.1186/s40168-017-0248-8.

[64] Azad MB, Bridgman SL, Becker AB, Kozyrskyj AL. Infant antibiotic exposure and the development of childhood overweight and central adiposity. Int J Obes (Lond) 2014;38:1290–8. Available from: https://doi.org/10.1038/ijo.2014.119.

[65] Saari A, Virta LJ, Sankilampi U, Dunkel L, Saxen H. Antibiotic exposure in infancy and risk of being overweight in the first 24 months of life. Pediatrics 2015;135:617–26. Available from: https://doi.org/10.1542/peds.2014-3407.

[66] Alm B, Erdes L, Möllborg P, Pettersson R, Norvenius SG, Aberg N, et al. Neonatal antibiotic treatment is a risk factor for early wheezing. Pediatrics 2008;121:697–702. Available from: https://doi.org/10.1542/peds.2007-1232.

[67] Santos VS, Freire MS, Santana RNS, Martins-Filho PRS, Cuevas LE, Gurgel RQ. Association between histamine-2 receptor antagonists and adverse outcomes in neonates: a systematic review and meta-analysis. PLoS One 2019;14:e0214135. Available from: https://doi.org/10.1371/journal.pone.0214135.

[68] Imhann F, Vich Vila A, Bonder MJ, Lopez Manosalva AG, Koonen DPY, Fu J, et al. The influence of proton pump inhibitors and other commonly used medication on the gut microbiota. Gut Microbes 2017;8:351–8. Available from: https://doi.org/10.1080/19490976.2017.1284732.

[69] Freedberg DE, Lamousé-Smith ES, Lightdale JR, Jin Z, Yang Y-X, Abrams JA. Use of acid suppression medication is associated with risk for *C. difficile* infection in infants and children: a population-based study. Clin Infect Dis 2015;61:912–17. Available from: https://doi.org/10.1093/cid/civ432.

[70] Castellani C, Singer G, Kashofer K, Huber-Zeyringer A, Flucher C, Kaiser M, et al. The influence of proton pump inhibitors on the fecal microbiome of infants with gastroesophageal reflux—a prospective longitudinal interventional study. Front Cell Infect Microbiol 2017;7:444. Available from: https://doi.org/10.3389/fcimb.2017.00444.

[71] Gupta RW, Tran L, Norori J, Ferris MJ, Eren AM, Taylor CM, et al. Histamine-2 receptor blockers alter the fecal microbiota in premature infants. J Pediatr Gastroenterol Nutr 2013;56:397–400. Available from: https://doi.org/10.1097/MPG.0b013e318282a8c2.

[72] Bokulich NA, Mills DA, Underwood MA. Surface microbes in the neonatal intensive care unit: changes with routine cleaning and over time. J Clin Microbiol 2013;51:2617–24. Available from: https://doi.org/10.1128/JCM.00898-13.

[73] Carl MA, Ndao IM, Springman AC, Manning SD, Johnson JR, Johnston BD, et al. Sepsis from the gut: the enteric habitat of bacteria that cause late-onset neonatal bloodstream infections. Clin Infect Dis 2014;58:1211–18. Available from: https://doi.org/10.1093/cid/ciu084.

[74] Neu J, Walker WA. Necrotizing enterocolitis. N Engl J Med 2011;364:255–64. Available from: https://doi.org/10.1056/NEJMra1005408.

[75] González-Rivera R, Culverhouse RC, Hamvas A, Tarr PI, Warner BB. The age of necrotizing enterocolitis onset: an application of Sartwell's incubation period model. J Perinatol 2011;31:519–23. Available from: https://doi.org/10.1038/jp.2010.193.

[76] Yazji I, Sodhi CP, Lee EK, Good M, Egan CE, Afrazi A, et al. Endothelial TLR4 activation impairs intestinal microcirculatory perfusion in necrotizing enterocolitis via eNOS–NO–nitrite signaling. Proc Natl Acad Sci U S A 2013;110:9451–6. Available from: https://doi.org/10.1073/pnas.1219997110.

[77] Claud EC, Lu L, Anton PM, Savidge T, Walker WA, Cherayil BJ. Developmentally regulated IkappaB expression in intestinal epithelium and susceptibility to flagellin-induced inflammation. Proc Natl Acad Sci U S A 2004;101:7404–8. Available from: https://doi.org/10.1073/pnas.0401710101.

[78] Hodzic Z, Bolock AM, Good M. The role of mucosal immunity in the pathogenesis of necrotizing enterocolitis. Front Pediatr 2017;5. Available from: https://doi.org/10.3389/fped.2017.00040.

[79] Egan CE, Sodhi CP, Good M, Lin J, Jia H, Yamaguchi Y, et al. Toll-like receptor 4-mediated lymphocyte influx induces neonatal necrotizing enterocolitis. J Clin Invest 2016;126:495–508. Available from: https://doi.org/10.1172/JCI83356.

[80] Bavineni M, Wassenaar TM, Agnihotri K, Ussery DW, Lüscher TF, Mehta JL. Mechanisms linking preterm birth to onset of cardiovascular disease later in adulthood. Eur Heart J 2019;40:1107–12. Available from: https://doi.org/10.1093/eurheartj/ehz025.

[81] White JR, Gong H, Pope B, Schlievert P, McElroy SJ. Paneth-cell-disruption-induced necrotizing enterocolitis in mice requires live bacteria and occurs independently of TLR4 signaling. Dis Model Mech 2017;10:727–36. Available from: https://doi.org/10.1242/dmm.028589.

[82] Morowitz MJ, Poroyko V, Caplan M, Alverdy J, Liu DC. Redefining the role of intestinal microbes in the pathogenesis of necrotizing enterocolitis. Pediatrics 2010;125:777–85. Available from: https://doi.org/10.1542/peds.2009-3149.

[83] Raveh-Sadka T, Thomas BC, Singh A, Firek B, Brooks B, Castelle CJ, et al. Gut bacteria are rarely shared by co-hospitalized premature infants, regardless of necrotizing enterocolitis development. Elife 2015;4. Available from: https://doi.org/10.7554/eLife.05477.

[84] Bergmann KR, Liu SXL, Tian R, Kushnir A, Turner JR, Li H-L, et al. Bifidobacteria stabilize claudins at tight junctions and prevent intestinal barrier dysfunction in mouse necrotizing enterocolitis. Am J Pathol 2013;182:1595–606. Available from: https://doi.org/10.1016/j.ajpath.2013.01.013.

[85] Wandro S, Osborne S, Enriquez C, Bixby C, Arrieta A, Whiteson K. The microbiome and metabolome of preterm infant stool are personalized and not driven by health outcomes, including necrotizing enterocolitis and late-onset sepsis. mSphere 2018;3. Available from: https://doi.org/10.1128/mSphere.00104-18.

[86] Normann E, Fahlén A, Engstrand L, Lilja HE. Intestinal microbial profiles in extremely preterm infants with and without necrotizing enterocolitis. Acta Paediatr 2013;102:129–36. Available from: https://doi.org/10.1111/apa.12059.

[87] Liu J, Li Y, Feng Y, Pan L, Xie Z, Yan Z, et al. Patterned progression of gut microbiota associated with necrotizing enterocolitis and late onset sepsis in preterm infants: a prospective study in a Chinese neonatal intensive care unit. PeerJ 2019;7:e7310. Available from: https://doi.org/10.7717/peerj.7310.

[88] Mai V, Young CM, Ukhanova M, Wang X, Sun Y, Casella G, et al. Fecal microbiota in premature infants prior to necrotizing enterocolitis. PLoS One 2011;6:e20647. Available from: https://doi.org/10.1371/journal.pone.0020647.

[89] Torrazza RM, Ukhanova M, Wang X, Sharma R, Hudak ML, Neu J, et al. Intestinal microbial ecology and environmental factors affecting necrotizing enterocolitis. PLoS One 2013;8:e83304. Available from: https://doi.org/10.1371/journal.pone.0083304.

[90] Morrow AL, Lagomarcino AJ, Schibler KR, Taft DH, Yu Z, Wang B, et al. Early microbial and metabolomic signatures predict later onset of necrotizing enterocolitis in preterm infants. Microbiome 2013;1:13. Available from: https://doi.org/10.1186/2049-2618-1-13.

[91] Taft DH, Ambalavanan N, Schibler KR, Yu Z, Newburg DS, Ward DV, et al. Intestinal microbiota of preterm infants differ over time and between hospitals. Microbiome 2014;2:36. Available from: https://doi.org/10.1186/2049-2618-2-36.

[92] Shaw AG, Sim K, Randell P, Cox MJ, McClure ZE, Li M-S, et al. Late-onset bloodstream infection and perturbed maturation of the gastrointestinal microbiota in premature infants. PLoS One 2015;10:e0132923. Available from: https://doi.org/10.1371/journal.pone.0132923.

[93] Madan JC, Salari RC, Saxena D, Davidson L, O'Toole GA, Moore JH, et al. Gut microbial colonisation in premature neonates predicts neonatal sepsis. Arch Dis Child Fetal Neonatal Ed 2012;97:F456–62. Available from: https://doi.org/10.1136/fetalneonatal-2011-301373.

[94] van den Berg JP, Westerbeek EAM, van der Klis FRM, Berbers GAM, van Elburg RM. Transplacental transport of IgG antibodies to preterm infants: a review of the literature. Early Hum Dev 2011;87:67–72. Available from: https://doi.org/10.1016/j.earlhumdev.2010.11.003.

[95] Smith MI, Yatsunenko T, Manary MJ, Trehan I, Mkakosya R, Cheng J, et al. Gut microbiomes of Malawian twin pairs discordant for kwashiorkor. Science 2013;339:548–54. Available from: https://doi.org/10.1126/science.1229000.

[96] Blanton LV, Charbonneau MR, Salih T, Barratt MJ, Venkatesh S, Ilkaveya O, et al. Gut bacteria that prevent growth impairments transmitted by microbiota from malnourished children. Science 2016;351. Available from: https://doi.org/10.1126/science.aad3311.

[97] Lu L, Yu Y, Guo Y, Wang Y, Chang EB, Claud EC. Transcriptional modulation of intestinal innate defense/inflammation genes by preterm infant microbiota in a humanized gnotobiotic mouse model. PLoS One 2015;10:e0124504. Available from: https://doi.org/10.1371/journal.pone.0124504.

[98] Arboleya S, Martinez-Camblor P, Solís G, Suárez M, Fernández N, de Los Reyes-Gavilán CG, et al. Intestinal microbiota and weight-gain in preterm neonates. Front Microbiol 2017;8:183. Available from: https://doi.org/10.3389/fmicb.2017.00183.

[99] Younge NE, Newgard CB, Cotten CM, Goldberg RN, Muehlbauer MJ, Bain JR, et al. Disrupted maturation of the microbiota and metabolome among extremely preterm infants with postnatal growth failure. Sci Rep 2019;9:8167. Available from: https://doi.org/10.1038/s41598-019-44547-y.

[100] Vael C, Verhulst SL, Nelen V, Goossens H, Desager KN. Intestinal microflora and body mass index during the first three years of life: an observational study. Gut Pathog 2011;3:8. Available from: https://doi.org/10.1186/1757-4749-3-8.

[101] Bergström A, Skov TH, Bahl MI, Roager HM, Christensen LB, Ejlerskov KT, et al. Establishment of intestinal microbiota during early life: a longitudinal, explorative study of a large cohort of Danish infants. Appl Env Microbiol 2014;80:2889–900. Available from: https://doi.org/10.1128/AEM.00342-14.

[102] Dogra S, Sakwinska O, Soh S-E, Ngom-Bru C, Brück WM, Berger B, et al. Dynamics of infant gut microbiota are influenced by delivery mode and gestational duration and are associated with subsequent adiposity. MBio 2015;6. Available from: https://doi.org/10.1128/mBio.02419-14.

[103] Kalliomäki M, Collado MC, Salminen S, Isolauri E. Early differences in fecal microbiota composition in children may predict overweight. Am J Clin Nutr 2008;87:534–8. Available from: https://doi.org/10.1093/ajcn/87.3.534.

[104] Forssberg H. Microbiome programming of brain development: implications for neurodevelopmental disorders. Dev Med Child Neurol 2019;61:744–9. Available from: https://doi.org/10.1111/dmcn.14208.

[105] Buffington SA, Di Prisco GV, Auchtung TA, Ajami NJ, Petrosino JF, Costa-Mattioli M. Microbial reconstitution reverses maternal diet-induced social and synaptic deficits in offspring. Cell 2016;165:1762–75. Available from: https://doi.org/10.1016/j.cell.2016.06.001.

[106] Diaz Heijtz R, Wang S, Anuar F, Qian Y, Björkholm B, Samuelsson A, et al. Normal gut microbiota modulates brain development and behavior. Proc Natl Acad Sci U S A 2011;108:3047–52. Available from: https://doi.org/10.1073/pnas.1010529108.

[107] Lu J, Synowiec S, Lu L, Yu Y, Bretherick T, Takada S, et al. Microbiota influence the development of the brain and behaviors in C57BL/6J mice. PLoS One 2018;13:e0201829. Available from: https://doi.org/10.1371/journal.pone.0201829.

[108] Lu J, Lu L, Yu Y, Cluette-Brown J, Martin CR, Claud EC. Effects of intestinal microbiota on brain development in humanized gnotobiotic mice. Sci Rep 2018;8:5443. Available from: https://doi.org/10.1038/s41598-018-23692-w.

[109] Kundu P, Blacher E, Elinav E, Pettersson S. Our gut microbiome: the evolving inner self. Cell 2017;171:1481–93. Available from: https://doi.org/10.1016/j.cell.2017.11.024.

[110] Lu J, Claud EC. Connection between gut microbiome and brain development in preterm infants. Dev Psychobiol 2018;. Available from: https://doi.org/10.1002/dev.21806.

[111] Niemarkt HJ, De Meij TG, van Ganzewinkel C, de Boer NKH, Andriessen P, Hütten MC, et al. Necrotizing enterocolitis, gut microbiota, and brain development: role of the brain-gut axis. Neonatology 2019;115:423–31. Available from: https://doi.org/10.1159/000497420.

[112] Bhutta AT, Cleves MA, Casey PH, Cradock MM, Anand KJS. Cognitive and behavioral outcomes of school-aged children who were born preterm: a meta-analysis. JAMA 2002;288:728–37. Available from: https://doi.org/10.1001/jama.288.6.728.

[113] Luu TM, Rehman Mian MO, Nuyt AM. Long-term impact of preterm birth: neurodevelopmental and physical health outcomes. Clin Perinatol 2017;44:305–14. Available from: https://doi.org/10.1016/j.clp.2017.01.003.

[114] You J, Shamsi BH, Hao M-C, Cao C-H, Yang W-Y. A study on the neurodevelopment outcomes of late preterm infants. BMC Neurol 2019;19:108. Available from: https://doi.org/10.1186/s12883-019-1336-0.

[115] Patrianakos-Hoobler AI, Msall ME, Huo D, Marks JD, Plesha-Troyke S, Schreiber MD. Predicting school readiness from neurodevelopmental assessments at age 2 years after respiratory distress syndrome in infants born preterm. Dev Med Child Neurol 2010;52:379–85. Available from: https://doi.org/10.1111/j.1469-8749.2009.03343.x.

[116] Schmidt B, Roberts RS, Davis PG, Doyle LW, Asztalos EV, Opie G, et al. Prediction of late death or disability at age 5 years using a count of 3 neonatal morbidities in very low birth weight infants. J Pediatr 2015;167:982–6. Available from: https://doi.org/10.1016/j.jpeds.2015.07.067 e2.

[117] Aylward GP. Neurodevelopmental outcomes of infants born prematurely. J Dev Behav Pediatr 2005;26:427.

[118] Sharon G, Sampson TR, Geschwind DH, Mazmanian SK. The central nervous system and the gut microbiome. Cell 2016;167:915–32. Available from: https://doi.org/10.1016/j.cell.2016.10.027.

[119] Hill C, Guarner F, Reid G, Gibson GR, Merenstein DJ, Pot B, et al. Expert consensus document. The International Scientific Association for Probiotics and Prebiotics consensus statement on the scope and appropriate use of the term probiotic. Nat Rev Gastroenterol Hepatol 2014;11:506–14. Available from: https://doi.org/10.1038/nrgastro.2014.66.

[120] Tanaka K, Nakamura Y, Terahara M, Yanagi T, Nakahara S, Furukawa O, et al. Poor bifidobacterial colonization is associated with late provision of colostrum and improved with probiotic supplementation in low birth weight infants. Nutrients 2019;11. Available from: https://doi.org/10.3390/nu11040839.

[121] Schultz M, Göttl C, Young RJ, Iwen P, Vanderhoof JA. Administration of oral probiotic bacteria to pregnant women causes temporary infantile colonization. J Pediatr Gastroenterol Nutr 2004;38:293–7. Available from: https://doi.org/10.1097/00005176-200403000-00012.

[122] Abdulkadir B, Nelson A, Skeath T, Marrs ECL, Perry JD, Cummings SP, et al. Routine use of probiotics in preterm infants: longitudinal impact on the microbiome and metabolome. Neonatology 2016;109:239–47. Available from: https://doi.org/10.1159/000442936.

[123] Dong Y, Speer CP. Late-onset neonatal sepsis: recent developments. Arch Dis Child Fetal Neonatal Ed 2015;100:F257–63. Available from: https://doi.org/10.1136/archdischild-2014-306213.

[124] AlFaleh K, Anabrees J. Probiotics for prevention of necrotizing enterocolitis in preterm infants. Cochrane Database Syst Rev 2014;CD005496. Available from: https://doi.org/10.1002/14651858.CD005496.pub4.

[125] Rao SC, Athalye-Jape GK, Deshpande GC, Simmer KN, Patole SK. Probiotic supplementation and late-onset sepsis in preterm infants: a meta-analysis. Pediatrics 2016;137:e20153684. Available from: https://doi.org/10.1542/peds.2015-3684.

[126] Yazdankhah SP, Narvhus JA, Eggesbø M, Midtvedt T, Lassen J, Halvorsen R. Should long-term prophylactic use of probiotics for infants and young children give cause for concern? Microb Ecol Health Dis 2008;20:171–6. Available from: https://doi.org/10.1080/08910600802504646.

[127] Claud EC. First do no harm. J Pediatr Pharmacol Ther 2012;17:298–301. Available from: https://doi.org/10.5863/1551-6776-17.4.298.

[128] Vallabhaneni S, Walker TA, Lockhart SR, Ng D, Chiller T, Melchreit R, et al. Fatal gastrointestinal mucormycosis in a premature infant associated with a contaminated dietary supplement—Connecticut, 2014. MMWR Morb Mortal Wkly Rep 2015;64:155–6.

[129] Luedtke SA, Yang JT, Wild HE. Probiotics and necrotizing enterocolitis: finding the missing pieces of the probiotic puzzle. J Pediatr Pharmacol Ther 2012;17:308–28. Available from: https://doi.org/10.5863/1551-6776-17.4.308.

[130] Ohishi A, Takahashi S, Ito Y, Ohishi Y, Tsukamoto K, Nanba Y, et al. *Bifidobacterium* septicemia associated with postoperative probiotic therapy in a neonate with omphalocele. J Pediatr 2010;156:679–81. Available from: https://doi.org/10.1016/j.jpeds.2009.11.041.

[131] Besselink MG, van Santvoort HC, Buskens E, Boermeester MA, van Goor H, Timmerman HM, et al. Probiotic prophylaxis in predicted severe acute pancreatitis: a randomised, double-blind, placebo-controlled trial. Lancet 2008;371:651–9. Available from: https://doi.org/10.1016/S0140-6736(08)60207-X.

[132] Gibson GR, Hutkins R, Sanders ME, Prescott SL, Reimer RA, Salminen SJ, et al. Expert consensus document: The International Scientific Association for Probiotics and Prebiotics (ISAPP) consensus statement on the definition and scope of prebiotics. Nat Rev Gastroenterol Hepatol 2017;14:491–502. Available from: https://doi.org/10.1038/nrgastro.2017.75.

[133] Pammi M, Suresh G. Enteral lactoferrin supplementation for prevention of sepsis and necrotizing enterocolitis in preterm infants. Cochrane Database Syst Rev 2017;6. Available from: https://doi.org/10.1002/14651858.CD007137.pub5.

[134] Srinivasjois R, Rao S, Patole S. Prebiotic supplementation in preterm neonates: updated systematic review and meta-analysis of randomised controlled trials. Clin Nutr 2013;32:958–65. Available from: https://doi.org/10.1016/j.clnu.2013.05.009.

[135] Underwood MA, Gaerlan S, De Leoz MLA, Dimapasoc L, Kalanetra KM, Lemay DG, et al. Human milk oligosaccharides in premature infants: absorption, excretion, and influence on the intestinal microbiota. Pediatr Res 2015;78:670–7. Available from: https://doi.org/10.1038/pr.2015.162.

[136] Parkinson JRC, Hyde MJ, Gale C, Santhakumaran S, Modi N. Preterm birth and the metabolic syndrome in adult life: a systematic review and meta-analysis. Pediatrics 2013;131:e1240–63. Available from: https://doi.org/10.1542/peds.2012-2177.

[137] Baraldi, E., Carraro, S., Filippone, M. Bronchopulmonary dysplasia: definitions and long-term respiratory outcome. Early Hum Dev 2009;85:S1–S3. In: Proceedings and selected abstracts from 1st international conference on clinical neonatology, Torino, Italy, November 2009. https://doi.org/10.1016/j.earlhumdev.2009.08.002.

Chapter 5

The developing microbiome and inflammatory bowel diseases

Stanley Cho[1], Mahesh Krishna[2] and Richard Kellermayer[1,3]
[1]*Section of Pediatric Gastroenterology, Texas Children's Hospital, Baylor College of Medicine, Houston, TX, United States,* [2]*Rice University, Houston, TX, United States,* [3]*USDA/ARS Children's Nutrition Research Center, Houston, TX, United States*

Introduction

In this chapter, we will discuss the current knowledge and controversies about the microbiome in inflammatory bowel diseases (IBDs), with an emphasis on how critical environmental influences on the developing microbiome may shape human susceptibility toward developing these diseases.

Inflammatory bowel diseases

IBDs are chronic, remitting–relapsing inflammatory conditions that primarily involve the gastrointestinal tract. Crohn, or Crohn's, disease (CD) and ulcerative colitis (UC) are the two main types of IBD, which are primarily differentiated by the location and extent of inflammation within the human gastrointestinal (GI) system [1]. UC mostly affects the colon, whereas CD can involve any part of the alimentary tract from the mouth to the anus, often in a discontinuous fashion. The inflammation in UC is generally limited to the mucosa, whereas inflammation in CD is commonly transmural. The clinical distinction between UC and CD can be challenging, however, especially in children [2]. In such instances a third classification of IBD, IBD-unclassified (IBD-U), is used. IBD-U defines patients with chronic indeterminate colitis who otherwise lack the distinguishing clinical or histologic features of UC or CD [1].

CD and UC are actually subdivided further based (i.e., Montreal and Paris classifications [3]) on age of presentation, localization of the intestinal inflammation, perianal manifestation, and the behavior of the inflammatory process (i.e., stricturing and or penetrating). The details of this subclassification extend

The Developing Microbiome. DOI: https://doi.org/10.1016/B978-0-12-820602-7.00005-2

beyond the scope of this chapter, but it is important to recognize that IBD is a diagnosis of exclusion, and thereby imperfect. Namely, the diagnosis of IBD in a patient is reached based on clinical history and presentation, supported by laboratory, imaging, and endoscopic findings, which aid in ruling out other diseases that can present similar to IBD, such as intestinal tuberculosis, for example. Even after classifying a patient according to current state of art, one must recognize that neither CD nor UC are single disorders, but rather groups of diseases, which we cannot at present delineate better [4]. Due to this heterogeneity, individual patients are likely to carry distinctive characteristics, that is, have a truly personalized disease.

IBD in the pediatric population presents unique challenges. Children with IBD are more likely to have extensive disease at the time of presentation and often follow a more severe disease course [5,6]. Young children and adolescents with IBD are particularly vulnerable to impaired growth and delays in puberty [7–10]. There are also differences in the disease location at presentation compared to adults [11,12].

As a chronic condition without a known medical cure, IBD is a significant cause of morbidity. This translates to annual healthcare costs that exceed $5.5 billion in the United States alone [13,14]. Understanding the mechanisms behind this disease will be crucial in our efforts to improving our prevention and treatment capabilities, and potentially identifying a cure.

Epidemiology

In the United States, it is estimated that 1 in every 200–300 people are affected by IBD [15]. Over much of the 20th century, industrialized or "Westernized" countries experienced significant rises in the incidence of IBD. In recent decades, as the growth of IBD in western/economically advanced counties has slowed, newly industrialized countries in Asia, Africa, and South America are now experiencing rapid rises in incidence [16]. Along with economic advancement propagating IBD [17], it has been observed that the descendants of immigrants from areas of low IBD prevalence to areas with high IBD prevalence rapidly assume the same risk of IBD as those people have who are endemic to the high IBD prevalence areas [18]. While industrialization has led to overall improvements in standards of living and sanitation, with a dramatic decline in infectious diseases, there have been unpredicted consequences of economic advancement including detrimental environmental exposures, which appear to have shaped the landscape of current autoimmune diseases, such as IBD (reviewed in [19,20]).

IBD can manifest at any age, but its incidence has been commonly observed to peak in young adulthood between 18 and 30 years of age with a second possible peak in the older adult and elderly population between 50 and 80 years of age [21,22]. There are variable observations regarding the incidence changes over recent years in respect to IBD and pediatric IBD, but the incidence in young (<10 years old) and very young (<5 years old, i.e., very early onset [VEO] IBD) children appears to continuously rise even in countries with the highest prevalence of the diseases [23].

It is also important to recognize that IBD tends to cluster in families, whereby in first-degree relatives of IBD patients, the relative risk of developing the same type of IBD ranges between 5 and 10 in the case of CD, and between 2 and 8 in case of UC. To complicate this issue, relatives of IBD patients can develop both forms of IBD, but CD patients have a higher frequency of relatives affected with UC when compared to UC patients having relatives affected with CD (reviewed in [24]). Such observations not only highlight the importance of genetics and/or close family environment, but also indicate that there are common features of both factors in regards to CD and UC pathogenesis, since both disorders can be present within the same family, bringing forth the concept of stochastic (i.e., random) biologic mechanisms at play.

Pathogenesis

The pathogenesis of IBD remains poorly understood. Many different factors have been implicated, and the current paradigm of IBD pathogenesis suggests that it is likely a multisystemic process influenced by many elements. Such systems and elements include genetic factors, environmental (including dietary) exposures, intestinal epithelium, host immune system, enteric nervous system, enteric vascular system, and the gut microbiome. The most prevailing current concept is that IBD develops as a result of environmentally influenced abnormal (i.e., autoimmune) immune activation against a component of the commensal microbiome in genetically predisposed individuals [25,26]. The exact mechanism for this autoimmune activation, however, is unknown and the definite pathways by which common phenotypes of IBD develop may actually vary between patients in an individualized and somewhat unpredictable (i.e., stochastic) manner (i.e., a truly personalized disease, reviewed in [27]).

A positive family history of IBD is the strongest risk factor for developing the disorders (as indicated before in the Epidemiology section), highlighting not only the role of genetics in pathogenesis, but also the potential role of closely shared (i.e., family shared) environmental exposures [24]. Namely, household-based environmental exposures and the possibility of IBD induced microbiome changes in a family member being shared with others, thereby increasing the chance of developing IBD within a family member cannot be easily dissected from genetics in such epidemiologic observations. First-degree relatives of those with IBD are about 3- to 20-fold more likely to develop the disorders than the general population [28,29]. In regards to genetics, genome wide association studies (GWAS) thus far have found more than

200 genomic loci to associate with IBD [30,31]. A significant portion of the affected genes is involved in the interactions between the gut immune system and the microbiome. For example, NOD2, the first gene where genetic polymorphisms were identified to increase risk for IBD, is known to stimulate an immune response upon recognition of the microbial cell wall component, muramyl dipeptide, of gram-positive and gram-negative bacteria. Patients with mutations of NOD2 have increased levels of mucosal adherent bacteria [32] as well as decreased levels of IL-10, an antiinflammatory cytokine [33]. Since the identification of NOD2 as a susceptibility locus, many other genetic loci-associated variation has been identified to confer risk for-, or protection against IBD.

Genetic predisposition to IBD—in particular, VEO-IBD [34]—is critical to pathogenesis, but it is recognized that genetic variation, such as single nucleotide polymorphisms (SNPs), alone cannot explain the observed heritability of IBD [35], nor its rising incidence over the past several decades [27,36]. Even in VEO-IBD, an IBD-prone polygenic background has been recognized [37], complicating our ability to clearly decipher IBD genomics.

Epigenetic modifications, which are mitotically heritable molecular changes that can modify gene expression without alteration of the genetic code, have also been suggested to be important in priming individuals for IBD [27,38–40]. Such modifications allow for adaptability in phenotypic expression in response to changing environmental stimuli. DNA cytosine methylation in CpG dinucleotides is perhaps the most studied example of an epigenetic modification that has been recognized as a potential pathogenic factor in IBD [27]. In a mouse model of IBD, for example, we have demonstrated that prenatal methyl-donor supplementation shifted postnatal colonic mucosal and cecal content microbiome composition in association with increased colitis susceptibility [41,42]. Such observations highlight the potential for prenatal environmental influences to result in longstanding nongenetic modifications (i.e., priming) in the mammalian epigenome, which can thereafter modulate susceptibility to IBD through altering critical pathogenic systems, such as the microbiome. Interestingly, components of methyl-donors are found in human prenatal supplements, which raises the question of shifting prenatal care practices adding to the increasing incidence of IBD in the developed world [41,42]. Indeed the maternal microbiome has been recognized to shape postnatal immune responses in offspring, most clearly demonstrated in imperfect animal models (reviewed in [43]). Such effects of the maternal microbiome on the prenatally (in utero) developing offspring are likely to be communicated by the meta-metabolome influencing fetal epigenetic development (Fig. 5.1). While prenatal microbial influences are predicted to be indirect (secondary to the generally sterile intrauterine environment), these observations point to the increasing attention towards the microbiome as pathogenic factor in autoimmune disorders, including IBD.

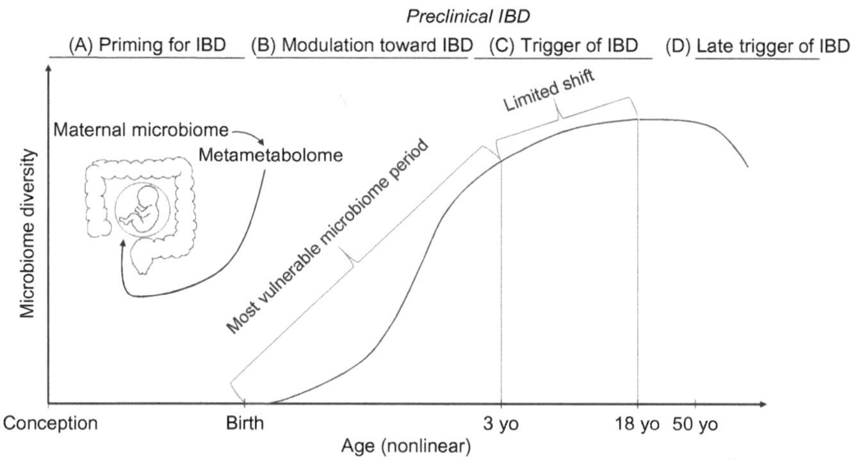

FIGURE 5.1 Graphical and theoretical representation of microbiome maturation and the relevant developmental window periods for inflammatory bowel disease (IBD) pathogenesis in humans. (A) The mostly sterile intrauterine development is critical for priming of the human host toward IBD. During this developmental period, the maternal microbiome may contribute to communicating detrimental environmental influences by the meta-metabolome toward the embryo, which increase the risk of postnatal IBD development, most likely through epigenetic changes (see chapter text). (B) After birth, there is a dramatic rise in gut microbiome diversity until round 2–3 years of age (yo), which creates the environmentally most vulnerable developmental window for the microbiome to modulate IBD susceptibility in the prenatally primed host. Under repeatedly harmful environmental stimulation, the priming and modulation of the individual (i.e,. host and commensal microbiome) can create a preclinical IBD (*italic*) state. (C). After 3 yo, the flux in the microbiome settles, but mild developmental maturation is likely to go on until young adulthood (18–30 yo); it is during this time period in which microbial and other indirect environmental triggers may bring about the onset of IBD (first peak of IBD incidence between 18 and 30 yo). (D) A second slow flux of decreasing microbiome diversity takes place following 50 yo during the elderly life period. It is this mild flux of the microbiome where microbial and other indirect environmental triggers may bring about the second rise of disease onset in the elderly people with prenatally primed and postnatally modulated preclinical IBD.

Gut microbiome in inflammatory bowel disease

Challenges in gut microbiome research

The commensal gut microbiota is the conglomerate of live microorganisms including bacteria, fungi, and viruses (less commonly protozoans and other parasites) that inhabit the gastrointestinal tract [44]. The gut microbiome includes the collection of genomes (i.e., metagenomes) and gene products (including RNA [metatranscriptome], proteins [metaproteome], and metabolites [meta-metabolome]) of the microbiota. It is a very diverse ecosystem with complex interactions. While the diversity of this microbial community has long been appreciated, we have been limited by our inability to accurately identify its components. With explosive advancement in genomic sequencing capabilities [45], our understanding of the gut microbiome's significance in human health and disease has shifted considerably. Along with this advancement, however, we have to recognize the nucleic acid-based interrogation and high-throughput technology-associated limitations, and consequent challenges for microbiome analytics. Such nucleic acid-based analytics is prone to bias from the point of sample selection and collection through storage, nucleic acid isolation, sequencing, and sequence identification to statistical analysis [46]. These limitations have naturally led to a translational research gap between basic science and bedside care with great efforts being taken to alleviate such gap [47]. Nevertheless, the true translational relevance of big data-based microbiome findings remain controversial in the clinical setting. While grandiose statements about the mammalian microbiome are frequently made, those can be seriously questioned. For example, it is frequently stated in publications and presentations that "the microbiome is essential for mammalian life" [48]. Actually, germ-free mice are the outstanding example to demonstrate that the microbiome is not critical for mammalian life. Rodents adapt well to this condition and to conventionalization as well [49]. Their immune system develops abnormally, but they actually live longer than conventionally raised pathogen-free animals [50]. Not only do germ-free mice live longer, but also they have less systemic inflammation and macrophage dysfunction than conventionally raised counterparts [51]. Clearly, much research needs to be done by dedicated collaborative physician scientists and basic researchers to identify key aspects of the mammalian microbiome, which may improve the clinical utilization of breakthrough basic scientific findings. This is rather important, since the current reality of targeted microbial therapeutics is rather dismal even in straightforward clinical applications such as acute gastroenteritis [52].

When considering challenges for gut microbiome research, we have to recognize that within the intestinal microbiome, fecal- and mucosa-associated communities significantly differ [53], and the mucosa-associated microbiome has more distinct disease associations than that of stool [54]. The majority of publications in the field, however, have not only been limited by studying fecal samples, but also by examining only the bacteriome at limited depth (16S rRNA gene-based sequencing) and in a trans-sectional (single time point) fashion. Yet, the importance of deeper level sequencing and assessing the function (i.e., metatranscriptome, meta-metabolome) rather than simply the genetic composition of the microbiome in a longitudinal (i.e., repeated sampling) fashion has been emphasized [55,56]. Furthermore, the importance of the fungome/mycobiome [57,58] and virome [59] (mostly bacteriophages) in shaping the intestinal microbiome is increasingly recognized, but much less studied than the bacteriome. Another recently addressed aspect of intestinal mucosal microbiome findings is that—at least in biopsy samples (the most frequent mode of sampling)—lamina propria phagocytes carry a separate "microbiome" [60]. Such compartmentalized "microbiome" is nonviable in the traditionally sense, and can be strongly argued as not even being part of the microbiome. This nonviable "microbiome", however, has been integral to intestinal biopsy-associated microbiome findings, which especially under inflammatory conditions (e.g., IBD) may be a significant component of microbial DNA content (i.e., the interrogated "microbiome"). Such cellular niche-related microbiome variation in the intestinal mucosa can also give rise to sampling method-related variation, as in the case of mucosal washings [61] and brushings [62] compared to biopsies.

We recognize that even the most careful and comprehensive approaches to studying the intestinal microbiome are limited by ethical, technical, and financial considerations, which commonly contribute to the challenges and bias listed in this section. In the meantime, it is very important to exercise careful criticism when interpreting current microbiome research findings with the aforementionede limitations in mind.

The "healthy" versus "dysbiotic" microbiome

For common disorders, such as IBD, where the gut microbiome is strongly indicated to play a role, it is increasingly becoming recognized that the interactive network of an "abnormal" microbiome/microbiota is a more likely pathogen, than a single microbe [54,63]. This abnormal state of a microbiome is designated as "dysbiosis" [63], which can associate with disease severity [54] and outcome [64,65]. Dysbiosis, however, can be difficult to clearly define due to the complexity of the microbiome, which leads to many challenges when trying to delineate the healthy state of that. Indeed, just as the healthy microbiome is influenced by geographical variation, age, diet, etc. [66], dysbiosis varies by disease states [67], significantly influenced by geographically dependent socioeconomic environments [68]. Longitudinal microbiome studies actually suggest the use of the "healthy plane" [56] instead of simply "healthy", underscoring the dynamic nature of the gut microbiome. Such healthy plane is characterized by greater richness and diversity compared to disease-associated gut microbiomes [44,66]. Certain bacterial taxa have been recognized to be enriched (i.e., protective) or decreased in abundance (i.e., harmful or deleterious) in the healthy plane. Specific bacteria, however, cannot be generally designated and "good" or "bad", however, but only in the context of a gut microbial community [69], and even beyond that, the health state of the host [70]. It is also important to note that bacterial diversity fluctuations within the healthy plane are modulated by their dynamic interaction with bacteriophages [71] and fungi [72] as well, the details of which are beyond the scope of this chapter.

Inflammatory bowel disease-related dysbiosis

In the context of IBD, specific components of an altered (i.e., dysbiotic, see previous section) microbiome plus other environmental factors in genetically susceptible individuals are likely to induce an abnormal autoimmune response that leads to the disease [73]. The IBD-associated dysbiosis is characterized by an intercalating increase in bacterial groups such as Proteobacteria, *Fusobacterium*, and *Ruminococcus*, as well as decreased amount of "protective" bacteria such as *Bifidobacterium* and *Roseburia* [73]. Murine models support the role of such dysbiosis in intestinal inflammation, since mice that have received their microbiota from feces of colitis-prone mice actually have more severe experimental colitis than those that have received stool from noncolitic animals [74].

There is notable evidence, however, that the relationship between the microbiome and the development of IBD is reciprocal, and that intestinal inflammation itself significantly alters the microbiome as well. From clinical observations and experimental studies, intestinal inflammation alters both bacterial composition and gene expression [73]. For example, intestinal inflammation has been shown to lead to a reduction in anaerobes, such as certain *Clostridium* groups, while causing an increase in aerobic and facultative anaerobic bacteria such as Proteobacteria via various proposed mechanisms including increased intraluminal oxygen concentration [53]. In addition, inflammation alters the function of the epithelial cells and their defense mechanisms, which in turn may explain the increase in adherent-invasive *Escherichia coli* seen in CD [75].

Due to the high variability in the composition of the gut microbiome between and within individuals [76], the elements that comprise a dysbiotic microbiome may significantly differ between IBD patients. Nonetheless, it has been observed that many of the bacteria that are correlated with IBD have specific roles in maintaining intestinal homeostasis. As reviewed by Sartor and Wu, Clostridia have been found to have immune-suppressive activity that protects against IBD by inducing regulatory T cells through a variety of mechanisms. This eventually leads to immunological tolerance against the gut microbiome, preventing intestinal inflammation as a result. *Faecalibacterium prausnitzii* is a bacterial species thought to protect against disease pathogenesis by stimulating IL-10 production from regulatory T cells as and by promoting protective metabolites such as salicylic acid, which protects against colitis [73].

In addition, there are bacteria that are correlated with IBD, which can disrupt intestinal homeostasis. Such an example is adherent and invasive *E. coli* (AIEC), found in almost 40% of patients with ileal CD, which can invade and replicate in epithelial cells, promoting expression of the inflammatory cytokine (IL-17) [77]. *Enterococcus faecalis*, which has been shown to disrupt epithelial barrier integrity in addition to activating innate immune pathways, is also an important species that has been associated with IBD development in addition to *Fusobacterium* spp. [78]. Cocolonization of *E. faecalis* with AIEC was found to significantly increase the severity and onset of experimental colitis, suggesting interaction [73].

Metabolic profiles can provide a glimpse into the pathogenesis of IBD as metabolites are the essential functional products of bacteria that have downstream effects on the immune system and beyond. Short-chain fatty acids (SCFA) are an example of a metabolite that plays an important role in mucosal protection and immune regulation. Patients with IBD are commonly found with significantly less SCFA, for example, consistent with the decreased numbers of butyrate-producing bacteria, such as *Roseburia*, Lachnospiraceae, and *F. prausnitzii* [79–81].

Animal models of IBD, however, are rather imperfect, and numerous observations even in humans give conflicting results when it comes to taxonomic observations. A great example for this is the bacterium *F. prausnitzii*, for example. There have been inconsistent results as to the role of *F. prausnitzii* in modulating intestinal inflammation, especially in UC have been observed [64].

As a summary for the gut microbiome in respect to IBD, it is important to repeatedly note that fecal- and mucosa-associated communities significantly differ [53], and specifically in IBD, the mucosa-associated microbiome has more distinct disease associations than that of stool [54]. In respect to the mucosa, lamina propria-associated mononuclear cell "microbiomes" have to be taken into consideration [60]. In spite of this, the examination of fecal samples is understandable, given ethical and clinical limitations of performing repeated invasive mucosal sampling that are required for longitudinal examinations [56]. Importantly, such longitudinal interrogations have highlighted that there can be significant disease state and patient-specific divergence between bacterial DNA-based abundance (metagenome) and transcriptional activity (metatranscriptome) [82]. This divergence, however, unfortunately questions the translational relevance of solely DNA-based microbiome observations in IBD. To further complicate our ability in deciphering the role of the microbiome in IBD, we have to highlight arguably the most significant undertaking to date in respect to studying the human microbiome [55]. In this study, even with guarded estimation, more than 2 million US dollars were spent on high-throughput laboratory investigations alone (not accounting for the cost of data and sample collection, or researcher salary). This large work concluded that interindividual variation exceeds any disease-associated measures in all aspects of the microbiome (metagenome, metatranscriptome, proteome, and metabolome) both in CD and UC. Furthermore, only 24.3% of samples in CD and 11.6% of samples in UC were determined to be dysbiotic. In respect to this dysbiosis, it was stated that "dysbiosis is triggered in part by stochastic (i.e., random, the authors) events" [55]. Taken together, based on existing evidence, a cause or effect role for the microbiome and IBD still cannot be conclusively be answered [83]. Therefore researches as well as clinicians must remain open-minded when it comes to the interpretation of data and the design of future studies.

The developing microbiome and inflammatory bowel disease

The commensal gut microbiome of humans changes with developmental stages of life (Fig. 5.1). The gut microbiome in early life is marked by high variability and low diversity, and it is more susceptible to fluctuations in response to environmental exposures and life events [84]. It is not until adulthood that the microbiome stabilizes and also reaches peak diversity [85]. The microbiome subsequently loses diversity and stability with aging beyond adulthood in the elderly, in conjunction with an increase in predisposition to inflammation [44,86,87]. Since IBD onset first peaks in young adulthood, we will focus mostly on infantile and pediatric microbiome development (Fig. 5.1).

Some studies indicated the existence of a perinatal/prenatal microbiome [88]; however, these findings could not be validated by recent direct observations indicating the lack of a placental microbiome [89]. Interestingly, antepartum and perinatal antibiotic exposure has been shown to augment colitis susceptibility later in life in mouse models [90,91], but this has not been supported by indirect human epidemiologic observations [92]. On the contrary, direct and especially repeated exposure to antibiotics by 2 years of age in humans has been observed to significantly increase the chances of subsequently developing IBD [93,94]. Therefore, we propose that current evidence indicates the human maternal microbiome to indirectly influence offspring susceptibility toward IBD by communicating dietary and other environmental influences through the meta-metabolome to the infant in utero (Fig. 5.1) as opposed to seeding a prenatal microbiome. Based on human migration studies, it is actually the intrauterine developmental period that is most critical for priming human susceptibility toward IBD [27]. Consequently, the postnatal microbiome is most likely to be only a modulator and perhaps a trigger for developing the diseases in the prenatally primed host (Fig. 5.1).

Following birth, the establishment of the microbiome in infants tends to occur in two major stages. The first stage is during lactation when an infant's gut microbiome is dominated by *Bifidobacterium*, though additional differences are noted between breast milk and formula-fed infants. This is followed by the introduction of solid foods (second stage), leading to a microbiome that begins to resemble that of an adult, marked by the presence of Bacteroidetes and Firmicutes [95] (see details in other chapters of this book). In infancy, studies have shown that birth route can affect the newborn microbiome. Vaginal delivery leads to a gut microbiome more heavily influenced by maternal vaginal and fecal flora such as *Lactobacilli*. Cesarean section on the other hand led to a gut microbiome more heavily influenced by skin flora (*Staphylococcus*, *Corynebacterium*, and *Propionibacterium*) [96,97]. Such delivery mode-associated offspring microbiome establishment is also modulated by maternal diet during pregnancy [98]. Thereafter, bacterial diversity rapidly increases throughout an infant's first few years in life along with decreasing interindividual variation across different ethnic populations around the world, with most of the microbiome being well established by 3–4 years of age [99].

The rapid changes in the microbiome that occur during early life have been shown, at least in mouse models, to lead to a vigorous immune response termed the "weaning response", an important element in the development of a healthy immune system [100]. In alignment with this observation, Olszak et al. [101] underscored that repopulation of germ-free mice with specific pathogen-free microbiota modulates intestinal mucosal immune responses and consequent susceptibility to colitis only in neonates, but not in adults. An elegant study using a germ-free mouse model demonstrated the critical role for bacterial colonization for intestinal epithelial cell maturation via epigenetic programming of crypt-based stem cells in mice as well [102]. Based on these observations in mammalian models, the microbiome-associated changes in epithelium, and mucosa-associated immune development, are likely to be restricted to early childhood (e.g., up to 3–4 years) highlighting the critical importance of this developmental time window. From the developmental origins of health and disease (DOHAD; https://dohadsoc.org/) standpoint, biological systems in flux/change are most vulnerable to environmental influences, which may leave a permanent predisposing alteration in a critical biologic system (the microbiome in this case) toward developing IBD later in life. Consequently, when we examine the role of the microbiome in the developmental origins of IBD, it is likely to play the most important role in the first 3–4 years of life in individuals. As already underscored, postnatal antibiotic exposure supports this proposition by indicating the first 2–3 years of infancy being most vulnerable to microbiome perturbations in regards to developing IBD later in life (Fig. 5.1). As for the postnatal epidemiologic examinations on early pediatric antibiotic exposure, however, the outcomes vary from study to study [103], but most are consistent for such association with CD as opposed to UC [104,105]. Naturally, many other environmental stimuli beyond antibiotics may induce critical changes in the microbiome during early postnatal pediatric development, which may modulate the IBD primed host toward "preclinical IBD" (Fig. 5.1). Just some examples from IBD models are early life stress [106], and maternal high-fat diet [107], but such exposures cannot be isolated to postnatal influences on the gut microbiome in humans (i.e., the exposure is not selective in the postnatal period, therefore leaving indirect embryonic effects of such stimuli influencing the developing epigenome a possibility, as opposed to influencing the postnatal microbiome in an isolated fashion, see earlier). Similar limitations exist in respect to passive smoking during early life (i.e., likely occurred during embryonic development as well) leading to a higher incidence of CD [108] and premature birth to higher incidence of both UC and CD [109] (i.e., unknown intrauterine influences leading to premature birth). In the meantime, mode of delivery (which significantly impacts initial microbiome establishment, see earlier) has not been observed to be a risk factor in human epidemiologic studies for developing IBD [110], while a systematic meta-analysis of the existing literature concluded that breastfeeding in infancy protects against the development of CD and UC [111].

To date, a limited number of studies have been performed in humans that focused on normal/control microbiome development during the pediatric age period beyond the first 3 years of life either in stool (i.e., fecal microbiome) [112] or colonic mucosa [113]. These studies, however implicate a significantly more subtle but still continued developmental change in the microbiome compared to that of infancy (Fig. 5.1). During this developmental period, critical changes in the microbiome may trigger the onset of disease in people with a "genetically originated," "prenatally primed" (through critical environmentally induced epigenetic changes), and "postnatally modulated" (perhaps through altered microbiome development in infancy) preclinical state of IBD. Indeed, a number of studies have observed IBD to be diagnosed following infections by gut pathogens (reviewed in [114]), or close to receiving antibiotics [115]. Both of these exposures directly disturb the intestinal microbiome. Other environmental perturbations such as academic stress [116] and/or geographic location shifts in early adulthood (transient emigration to countries with advanced economy) [117] are examples for environmental influences, which may indirectly modify host microbiome composition and trigger the onset of IBD.

Clinical implications

Based on previous discussion in this chapter, we propose that the prenatal maternal microbiome is an important communicator for critical IBD-prone environmental stimuli to prime the embryo (with IBD-prone genetic origins) toward developing IBD after birth. Following delivery, postnatal environmental changes negatively influencing early microbiome development can modulate the prenatally primed host toward a preclinical state of IBD (Fig. 5.1).

Identifying the critical pre- and postnatal environmental culprits can guide prevention against the disorders. One such example is antibiotic stewardship during infancy, which may significantly decrease the risk of IBD (especially CD) later in life.

Identifying the critical biologic system or factor that is primed by the prenatal, and modulated by the postnatal IBD-prone insults could point to the cure in an individualized fashion for a single patient. Such identification, however, is a tantalizing task taking into consideration the complexity of the biological systems involved: IBD-prone polygenome, immune system, gut mucosa, microbiome, etc. Therefore, we propose that the most feasible means for combating IBD moving forward is prevention from critical environmental culprits.

It is within the concept of a preclinical IBD state where commensal and/or pass-through microbes such as *Mycobacterium paratuberculosis* (MAP) [118], *Candida* [119,120], or *Malassezia* [58,121,122] may be triggers of the disease in select cases, which may provide targets primarily for therapy, perhaps for prevention, but rather unlikely for cure.

Conclusion/future directions/adulthood

While the microbiome is most dynamic during young pediatric age, and thus most likely to contribute to development of IBD during that period of life, it probably remains a key determinant of IBD onset (i.e., trigger, see Fig. 5.1C), course, and severity into adulthood and in the elderly as well. This is highlighted by the bimodal distribution of IBD onset with a late peak in older adults and the elderly between 50 and 80 years of age (see epidemiology and Fig. 5.1D: late trigger). Studies have demonstrated that aging beyond adulthood often leads to dysbiosis or progressive alteration in microbiome composition compared to younger gut microbiomes. This elderly flux has been associated with adverse outcomes including predisposition to intestinal inflammation [87,123]. As such, dysbiotic changes to the microbiome may be a potential mechanism by which age dependent environmental influences modulate susceptibility to IBD in infants and trigger the onset of IBD in young adults and the elderly (i.e., during slow shifts of the microbiome, see Fig. 5.1), emphasizing the importance of the microbiome to IBD at all stages of life. Therefore, a strategy to decipher the details of microbiome flux during infancy, later pediatric development, and aging beyond 50 years of life, along with unraveling the biological machinery governing such changes will be critical toward developing microbiome-based preventative and therapeutic measures against the increasing disease burden of IBD in human kind.

Acknowledgment

Richard Kellermayer was supported in part by philanthropic funds from the Wagner Family led Gutsy Kids Fund, and by the Klaasmeyer family funds for primary sclerosing colitis research.

References

[1] Tontini GE, Vecchi M, Pastorelli L, Neurath MF, Neumann H. Differential diagnosis in inflammatory bowel disease colitis: state of the art and future perspectives. World J Gastroenterol 2015;21(1):21–46.

[2] Queliza K, Ihekweazu FD, Schady D, Jensen C, Kellermayer R. Granulomatous upper gastrointestinal inflammation in pediatric ulcerative colitis. J Pediatr Gastroenterol Nutr 2018;66(4):620–3.

[3] Levine A, Griffiths A, Markowitz J, et al. Pediatric modification of the Montreal classification for inflammatory bowel disease: the Paris classification. Inflamm Bowel Dis 2011;17(6):1314–21.

[4] Kumar M, Garand M, Al Khodor S. Integrating omics for a better understanding of inflammatory bowel disease: a step towards personalized medicine. J Transl Med 2019;17(1):419.

[5] Jakobsen C, Bartek J, Wewer V, et al. Differences in phenotype and disease course in adult and paediatric inflammatory bowel disease—a population-based study. Aliment Pharmacol Ther 2011;34(10):1217–24.

[6] Langholz E, Munkholm P, Krasilnikoff PA, Binder V. Inflammatory bowel diseases with onset in childhood. Clinical features, morbidity, and mortality in a regional cohort. Scand J Gastroenterol 1997;32(2):139–47.

[7] Pfefferkorn M, Burke G, Griffiths A, et al. Growth abnormalities persist in newly diagnosed children with crohn disease despite current treatment paradigms. J Pediatr Gastroenterol Nutr 2009;48(2):168–74.

[8] Griffiths AM, Nguyen P, Smith C, MacMillan JH, Sherman PM. Growth and clinical course of children with Crohn's disease. Gut 1993;34(7):939–43.

[9] Hildebrand H, Karlberg J, Kristiansson B. Longitudinal growth in children and adolescents with inflammatory bowel disease. J Pediatr Gastroenterol Nutr 1994;18(2):165–73.

[10] Kanof ME, Lake AM, Bayless TM. Decreased height velocity in children and adolescents before the diagnosis of Crohn's disease. Gastroenterology 1988;95(6):1523–7.

[11] Kugathasan S, Judd RH, Hoffmann RG, et al. Epidemiologic and clinical characteristics of children with newly diagnosed inflammatory bowel disease in Wisconsin: a statewide population-based study. J Pediatr 2003;143(4):525–31.

[12] Rosen MJ, Dhawan A, Saeed SA. Inflammatory bowel disease in children and adolescents. JAMA Pediatr 2015;169(11):1053–60.

[13] Park KT, Bass D. Inflammatory bowel disease-attributable costs and cost-effective strategies in the United States: a review. Inflamm Bowel Dis 2011;17(7):1603–9.

[14] Park MD, Bhattacharya J, Park K. Differences in healthcare expenditures for inflammatory bowel disease by insurance status, income, and clinical care setting. PeerJ 2014;2:e587.

[15] Ng SC, Shi HY, Hamidi N, et al. Worldwide incidence and prevalence of inflammatory bowel disease in the 21st century: a systematic review of population-based studies. Lancet 2018;390(10114):2769–78.

[16] Molodecky NA, Soon IS, Rabi DM, et al. Increasing incidence and prevalence of the inflammatory bowel diseases with time, based on systematic review. Gastroenterology 2012;142(1):46–54 e42; quiz e30.

[17] Krishna M, Salako A, Fofanova T, Kellermayer R. Parental education may differentially impact pediatric inflammatory bowel disease phenotype risk. Inflamm Bowel Dis 2019;.

[18] Ponder A, Long MD. A clinical review of recent findings in the epidemiology of inflammatory bowel disease. Clin Epidemiol 2013;5:237–47.

[19] Fofanova TY, Petrosino JF, Kellermayer R. Microbiome-epigenome interactions and the environmental origins of inflammatory bowel diseases. J Pediatr Gastroenterol Nutr 2016;62(2):208–19.

[20] Bach JF. The effect of infections on susceptibility to autoimmune and allergic diseases. N Engl J Med 2002;347(12):911–20.

[21] Keyashian K, Dehghan M, Sceats L, Kin C, Limketkai BN, Park KT. Comparative incidence of inflammatory bowel disease in different age groups in the United States. Inflamm Bowel Dis 2019;25(12):1983–9.

[22] Shivashankar R, Tremaine WJ, Harmsen WS, Loftus EV. Incidence and prevalence of Crohn's disease and ulcerative colitis in Olmsted County, Minnesota from 1970 through 2010. Clin Gastroenterol Hepatol 2017;15(6):857–63.

[23] Benchimol EI, Mack DR, Nguyen GC, et al. Incidence, outcomes, and health services burden of very early onset inflammatory bowel disease. Gastroenterology 2014;147(4):803–13 e807; quiz e814-805.

[24] Nunes T, Fiorino G, Danese S, Sans M. Familial aggregation in inflammatory bowel disease: is it genes or environment? World J Gastroenterol 2011;17(22):2715–22.

[25] Sartor RB. Mechanisms of disease: pathogenesis of Crohn's disease and ulcerative colitis. Nat Clin Pract Gastroenterol Hepatol 2006;3(7):390–407.

[26] Kellermayer R. Genetic drift. "Omics"as the filtering gateway between environment and phenotype: The inflammatory bowel diseases example. Am J Med Genet A 2010;152A (12):3022–5.

[27] Kellermayer R. Challenges for epigenetic research in inflammatory bowel diseases. Epigenomics 2017;9(4):527–38.

[28] Satsangi J, Grootscholten C, Holt H, Jewell DP. Clinical patterns of familial inflammatory bowel disease. Gut 1996;38(5):738–41.

[29] Yang H, McElree C, Roth MP, Shanahan F, Targan SR, Rotter JI. Familial empirical risks for inflammatory bowel disease: differences between Jews and non-Jews. Gut 1993;34 (4):517–24.

[30] Ye BD, McGovern DP. Genetic variation in IBD: progress, clues to pathogenesis and possible clinical utility. Expert Rev Clin Immunol 2016;12(10):1091–107.

[31] Liu JZ, van Sommeren S, Huang H, et al. Association analyses identify 38 susceptibility loci for inflammatory bowel disease and highlight shared genetic risk across populations. Nat Genet 2015;47(9):979–86.

[32] Swidsinski A, Ladhoff A, Pernthaler A, et al. Mucosal flora in inflammatory bowel disease. Gastroenterology 2002;122(1):44–54.

[33] Philpott DJ, Girardin SE. Crohn's disease-associated Nod2 mutants reduce IL10 transcription. Nat Immunol 2009;10(5):455–7.

[34] Kelsen JR, Sullivan KE, Rabizadeh S, et al. North American Society for pediatric gastroenterology, hepatology, and nutrition position paper on the evaluation and management for patients with very early-onset inflammatory bowel disease. J Pediatr Gastroenterol Nutr 2020;70(3):389–403.

[35] Turpin W, Goethel A, Bedrani L, Croitoru Mdcm K. Determinants of IBD heritability: genes, bugs, and more. Inflamm Bowel Dis 2018;24(6):1133–48.

[36] Kellermayer R. Epigenetics and the developmental origins of inflammatory bowel diseases. Can J Gastroenterol 2012;26(12):909–15.

[37] Serra EG, Schwerd T, Moutsianas L, et al. Somatic mosaicism and common genetic variation contribute to the risk of very-early-onset inflammatory bowel disease. Nat Commun 2020;11(1):995.

[38] Zilbauer M, Zellos A, Heuschkel R, et al. Epigenetics in paediatric gastroenterology, hepatology, and nutrition: present trends and future perspectives. J Pediatr Gastroenterol Nutr 2016;62(4):521–9.

[39] Kraiczy J, Nayak K, Ross A, et al. Assessing DNA methylation in the developing human intestinal epithelium: potential link to inflammatory bowel disease. Mucosal Immunol 2016;9(3):647–58.

[40] Harris RA, Nagy-Szakal D, Kellermayer R. Human metastable epiallele candidates link to common disorders. Epigenetics 2013;8(2):157–63.

[41] Schaible TD, Harris RA, Dowd SE, Smith CW, Kellermayer R. Maternal methyl-donor supplementation induces prolonged murine offspring colitis susceptibility in association with mucosal epigenetic and microbiomic changes. Hum Mol Genet 2011;20(9):1687–96.

[42] Mir SA, Nagy-Szakal D, Dowd SE, Szigeti RG, Smith CW, Kellermayer R. Prenatal methyl-donor supplementation augments colitis in young adult mice. PLoS One 2013;8 (8):e73162.

[43] Nyangahu DD, Jaspan HB. Influence of maternal microbiota during pregnancy on infant immunity. Clin Exp Immunol 2019;198(1):47–56.

[44] Hollister EB, Gao C, Versalovic J. Compositional and functional features of the gastrointestinal microbiome and their effects on human health. Gastroenterology 2014;146 (6):1449–58.

[45] Waldor MK, Tyson G, Borenstein E, et al. Where next for microbiome research? PLoS Biol 2015;13(1):e1002050.

[46] Tyler AD, Smith MI, Silverberg MS. Analyzing the human microbiome: a "how to" guide for physicians. Am J Gastroenterol 2014;109(7):983–93.

[47] Kong HH, Segre JA. Bridging the translational research gap: a successful partnership involving a physician and a basic scientist. J Invest Dermatol 2010;130(6):1478–80.

[48] McKenzie VJ, Song SJ, Delsuc F, et al. The effects of captivity on the mammalian gut microbiome. Integr Comp Biol 2017;57(4):690–704.

[49] El Aidy S, van Baarlen P, Derrien M, et al. Temporal and spatial interplay of microbiota and intestinal mucosa drive establishment of immune homeostasis in conventionalized mice. Mucosal Immunol 2012;5(5):567–79.

[50] Tazume S, Umehara K, Matsuzawa H, Aikawa H, Hashimoto K, Sasaki S. Effects of germfree status and food restriction on longevity and growth of mice. Jikken Dobutsu 1991;40(4):517–22.

[51] Thevaranjan N, Puchta A, Schulz C, et al. Age-associated microbial dysbiosis promotes intestinal permeability, systemic inflammation, and macrophage dysfunction. Cell Host Microbe 2017;21(4):455–66 e454.

[52] Schnadower D, Tarr PI, Casper TC, et al. *Lactobacillus rhamnosus* GG versus placebo for acute eastroenteritis in children. N Engl J Med 2018;379(21):2002–14.

[53] Albenberg L, Esipova TV, Judge CP, et al. Correlation between intraluminal oxygen gradient and radial partitioning of intestinal microbiota. Gastroenterology 2014;147 (5):1055–63 e1058.

[54] Gevers D, Kugathasan S, Denson LA, et al. The treatment-naive microbiome in new-onset Crohn's disease. Cell Host Microbe 2014;15(3):382–92.

[55] Lloyd-Price J, Arze C, Ananthakrishnan AN, et al. Multi-omics of the gut microbial ecosystem in inflammatory bowel diseases. Nature 2019;569(7758):655–62.

[56] Halfvarson J, Brislawn CJ, Lamendella R, et al. Dynamics of the human gut microbiome in inflammatory bowel disease. Nat Microbiol 2017;2:17004.

[57] Sokol H, Leducq V, Aschard H, et al. Fungal microbiota dysbiosis in IBD. Gut 2017;66 (6):1039–48.

[58] Kellermayer R, Mir SA, Nagy-Szakal D, et al. Microbiota separation and C-reactive protein elevation in treatment-naive pediatric granulomatous Crohn disease. J Pediatr Gastroenterol Nutr 2012;55(3):243–50.

[59] Dzunkova M, Low SJ, Daly JN, Deng L, Rinke C, Hugenholtz P. Defining the human gut host-phage network through single-cell viral tagging. Nat Microbiol 2019;4 (12):2192–203.

[60] Dheer R, Davies JM, Quintero MA, et al. Microbial signatures and innate immune gene expression in lamina propria phagocytes of inflammatory bowel disease patients. Cell Mol Gastroenterol Hepatol 2019;9(3):387–402.

[61] Tong M, Jacobs JP, McHardy IH, Braun J. Sampling of intestinal microbiota and targeted amplification of bacterial 16S rRNA genes for microbial ecologic analysis. Curr Protoc Immunol 2014;107 7.41.1–7.41.11.

[62] Nishino K, Nishida A, Inoue R, et al. Analysis of endoscopic brush samples identified mucosa-associated dysbiosis in inflammatory bowel disease. J Gastroenterol 2018;53 (1):95–106.

[63] DeGruttola AK, Low D, Mizoguchi A, Mizoguchi E. Current understanding of dysbiosis in disease in human and animal models. Inflamm Bowel Dis 2016;22(5):1137–50.

[64] Shah R, Cope JL, Nagy-Szakal D, et al. Composition and function of the pediatric colonic mucosal microbiome in untreated patients with ulcerative colitis. Gut Microbes 2016;7(5):384–96.

[65] Kugathasan S, Denson LA, Walters TD, et al. Prediction of complicated disease course for children newly diagnosed with Crohn's disease: a multicentre inception cohort study. Lancet 2017;389(10080):1710–18.

[66] Ruan W, Engevik MA, Spinler JK, Versalovic J. Healthy human gastrointestinal microbiome: composition and function after a decade of exploration. Dig Dis Sci 2020;;65(3).

[67] Walters WA, Xu Z, Knight R. Meta-analyses of human gut microbes associated with obesity and IBD. FEBS Lett 2014;588(22):4223–33.

[68] He Y, Wu W, Zheng HM, et al. Regional variation limits applications of healthy gut microbiome reference ranges and disease models. Nat Med 2018;24(10):1532–5.

[69] Kellermayer R. Fecal microbiota transplantation: great potential with many challenges. Transl Gastroenterol Hepatol 2019;4:40.

[70] Kellermayer R. Burdening questions about *Clostridium difficile* in pediatric inflammatory bowel diseases. J Pediatr Gastroenterol Nutr 2015;60(4):421–2.

[71] Manrique P, Dills M, Young MJ. The human gut phage community and its implications for health and disease. Viruses 2017;9(6).

[72] Limon JJ, Skalski JH, Underhill DM. Commensal fungi in health and disease. Cell Host Microbe 2017;22(2):156–65.

[73] Sartor RB, Wu GD. Roles for intestinal bacteria, viruses, and fungi in pathogenesis of inflammatory bowel diseases and therapeutic approaches. Gastroenterology 2017;152(2):327–39 e324.

[74] Elinav E, Strowig T, Kau AL, et al. NLRP6 inflammasome regulates colonic microbial ecology and risk for colitis. Cell 2011;145(5):745–57.

[75] Darfeuille-Michaud A, Boudeau J, Bulois P, et al. High prevalence of adherent-invasive *Escherichia coli* associated with ileal mucosa in Crohn's disease. Gastroenterology 2004;127(2):412–21.

[76] Consortium HMP. Structure, function and diversity of the healthy human microbiome. Nature 2012;486(7402):207–14.

[77] Cespedes S, Saitz W, Del Canto F, et al. Genetic diversity and virulence determinants of *Escherichia coli* strains isolated from patients with Crohn's disease in Spain and Chile. Front Microbiol 2017;8:639.

[78] Zhou Y, Chen H, He H, et al. Increased *Enterococcus faecalis* infection is associated with clinically active Crohn disease. Med (Baltim) 2016;95(39):e5019.

[79] Sokol H, Pigneur B, Watterlot L, et al. *Faecalibacterium prausnitzii* is an anti-inflammatory commensal bacterium identified by gut microbiota analysis of Crohn disease patients. Proc Natl Acad Sci U S A 2008;105(43):16731–6.

[80] Frank DN, St Amand AL, Feldman RA, Boedeker EC, Harpaz N, Pace NR. Molecular-phylogenetic characterization of microbial community imbalances in human inflammatory bowel diseases. Proc Natl Acad Sci U S A 2007;104(34):13780–5.

[81] Machiels K, Joossens M, Sabino J, et al. A decrease of the butyrate-producing species *Roseburia hominis* and *Faecalibacterium prausnitzii* defines dysbiosis in patients with ulcerative colitis. Gut 2014;63(8):1275–83.

[82] Schirmer M, Franzosa EA, Lloyd-Price J, et al. Dynamics of metatranscription in the inflammatory bowel disease gut microbiome. Nat Microbiol 2018;3(3):337–46.

[83] Ni J, Wu GD, Albenberg L, Tomov VT. Gut microbiota and IBD: causation or correlation? Nat Rev Gastroenterol Hepatol 2017;14(10):573–84.

[84] Dominguez-Bello MG, Blaser MJ, Ley RE, Knight R. Development of the human gastrointestinal microbiota and insights from high-throughput sequencing. Gastroenterology 2011;140(6):1713–19.

[85] Lozupone CA, Stombaugh JI, Gordon JI, Jansson JK, Knight R. Diversity, stability and resilience of the human gut microbiota. Nature 2012;489(7415):220–30.

[86] Kostic AD, Xavier RJ, Gevers D. The microbiome in inflammatory bowel disease: current status and the future ahead. Gastroenterology 2014;146(6):1489–99.

[87] Claesson MJ, Jeffery IB, Conde S, et al. Gut microbiota composition correlates with diet and health in the elderly. Nature 2012;488(7410):178–84.

[88] Aagaard K, Ma J, Antony KM, Ganu R, Petrosino J, Versalovic J. The placenta harbors a unique microbiome. Sci Transl Med 2014;6(237):237–65.

[89] de Goffau MC, Lager S, Sovio U, et al. Human placenta has no microbiome but can contain potential pathogens. Nature 2019;572(7769):329–34.

[90] Miyoshi J, Bobe AM, Miyoshi S, et al. Peripartum antibiotics promote gut dysbiosis, loss of immune tolerance, and inflammatory bowel disease in genetically prone offspring. Cell Rep 2017;20(2):491–504.

[91] Munyaka PM, Eissa N, Bernstein CN, Khafipour E, Ghia JE. Antepartum antibiotic treatment increases offspring susceptibility to experimental colitis: a role of the gut microbiota. PLoS One 2015;10(11):e0142536.

[92] Bernstein CN, Burchill C, Targownik LE, Singh H, Ghia JE, Roos LL. Maternal infections that would warrant antibiotic use antepartum or peripartum are not a risk factor for the development of IBD: a population-based analysis. Inflamm Bowel Dis 2017;23(4):635–40.

[93] Kronman MP, Zaoutis TE, Haynes K, Feng R, Coffin SE. Antibiotic exposure and IBD development among children: a population-based cohort study. Pediatrics 2012;130(4):e794–803.

[94] Shaw SY, Blanchard JF, Bernstein CN. Association between the use of antibiotics in the first year of life and pediatric inflammatory bowel disease. Am J Gastroenterol 2010;105(12):2687–92.

[95] Tanaka M, Nakayama J. Development of the gut microbiota in infancy and its impact on health in later life. Allergol Int 2017;66(4):515–22.

[96] Dominguez-Bello MG, Costello EK, Contreras M, et al. Delivery mode shapes the acquisition and structure of the initial microbiota across multiple body habitats in newborns. Proc Natl Acad Sci U S A 2010;107(26):11971–5.

[97] Mackie RI, Sghir A, Gaskins HR. Developmental microbial ecology of the neonatal gastrointestinal tract. Am J Clin Nutr 1999;69(5):1035S–45S.

[98] Lundgren SN, Madan JC, Emond JA, et al. Maternal diet during pregnancy is related with the infant stool microbiome in a delivery mode-dependent manner. Microbiome 2018;6(1):109.

[99] Yatsunenko T, Rey FE, Manary MJ, et al. Human gut microbiome viewed across age and geography. Nature 2012;486(7402):222–7.

[100] Al Nabhani Z, Dulauroy S, Marques R, et al. A weaning reaction to microbiota is required for resistance to immunopathologies in the adult. Immunity 2019;50(5):1276–88 e1275.

[101] Olszak T, An D, Zeissig S, et al. Microbial exposure during early life has persistent effects on natural killer T cell function. Science 2012;336(6080):489–93.

[102] Yu DH, Gadkari M, Zhou Q, et al. Postnatal epigenetic regulation of intestinal stem cells requires DNA methylation and is guided by the microbiome. Genome Biol 2015;16:211.
[103] Troelsen FS, Jick S. Antibiotic use in childhood and adolescence and risk of inflammatory bowel disease: a case-control study in the UK clinical practice research datalink. Inflamm Bowel Dis 2019;.
[104] Virta L, Auvinen A, Helenius H, Huovinen P, Kolho KL. Association of repeated exposure to antibiotics with the development of pediatric Crohn's disease—a nationwide, register-based finnish case-control study. Am J Epidemiol 2012;175(8):775–84.
[105] Ungaro R, Bernstein CN, Gearry R, et al. Antibiotics associated with increased risk of new-onset Crohn's disease but not ulcerative colitis: a meta-analysis. Am J Gastroenterol 2014;109(11):1728–38.
[106] Lennon EM, Maharshak N, Elloumi H, Borst L, Plevy SE, Moeser AJ. Early life stress triggers persistent colonic barrier dysfunction and exacerbates colitis in adult IL-10-/- mice. Inflamm Bowel Dis 2013;19(4):712–19.
[107] Xie R, Sun Y, Wu J, et al. Maternal high fat diet alters gut microbiota of offspring and exacerbates DSS-induced colitis in adulthood. Front Immunol 2018;9:2608.
[108] Mahid SS, Minor KS, Stromberg AJ, Galandiuk S. Active and passive smoking in childhood is related to the development of inflammatory bowel disease. Inflamm Bowel Dis 2007;13(4):431–8.
[109] Sonntag B, Stolze B, Heinecke A, et al. Preterm birth but not mode of delivery is associated with an increased risk of developing inflammatory bowel disease later in life. Inflamm Bowel Dis 2007;13(11):1385–90.
[110] Bernstein CN, Banerjee A, Targownik LE, et al. Cesarean section delivery is not a risk factor for development of inflammatory bowel disease: a population-based analysis. Clin Gastroenterol Hepatol 2016;14(1):50–7.
[111] Xu L, Lochhead P, Ko Y, Claggett B, Leong RW, Ananthakrishnan AN. Systematic review with meta-analysis: breastfeeding and the risk of Crohn's disease and ulcerative colitis. Aliment Pharmacol Ther 2017;46(9):780–9.
[112] Hollister EB, Riehle K, Luna RA, et al. Structure and function of the healthy preadolescent pediatric gut microbiome. Microbiome 2015;3:36.
[113] Harris RA, Shah R, Hollister EB, et al. Colonic mucosal epigenome and microbiome development in children and adolescents. J Immunol Res 2016;2016:9170162.
[114] Axelrad JE, Joelson A, Green PHR, et al. Enteric infections are common in patients with flares of inflammatory bowel disease. Am J Gastroenterol 2018;113(10):1530–9.
[115] Shaw SY, Blanchard JF, Bernstein CN. Association between the use of antibiotics and new diagnoses of Crohn's disease and ulcerative colitis. Am J Gastroenterol 2011;106(12):2133–42.
[116] Krishna MZ, Barton KR, Perez CM, Walsh SM, Assa A, Kellermayer R. Academic stress may contribute to the onset of pediatric inflammatory bowel diseases. J Pediatr Gastroenterol Nutr 2018;67(4):e73–6.
[117] Barreiro-de Acosta M, Alvarez Castro A, Souto R, Iglesias M, Lorenzo A, Dominguez-Munoz JE. Emigration to western industrialized countries: a risk factor for developing inflammatory bowel disease. J Crohns Colitis 2011;5(6):566–9.
[118] Pierce ES, Borowitz SM, Naser SA. The broad street pump revisited: dairy farms and an ongoing outbreak of inflammatory bowel disease in forest, Virginia. Gut Pathog 2011;3(1):20.
[119] Standaert-Vitse A, Sendid B, Joossens M, et al. *Candida albicans* colonization and ASCA in familial Crohn's disease. Am J Gastroenterol 2009;104(7):1745–53.

[120] Hoarau G, Mukherjee PK, Gower-Rousseau C, et al. Bacteriome and mycobiome interactions underscore microbial dysbiosis in familial Crohn's Disease. MBio 2016;7(5).
[121] Limon JJ, Tang J, Li D, et al. Malassezia is associated with Crohn's disease and exacerbates colitis in mouse models. Cell Host Microbe 2019;25(3):377–88 e376.
[122] Laurence M, Asquith M, Rosenbaum JT. Spondyloarthritis, acute anterior uveitis, and fungi: updating the Catterall-King hypothesis. Front Med (Lausanne) 2018;5:80.
[123] Maffei VJ, Kim S, Blanchard E, et al. Biological aging and the human gut microbiota. J Gerontol A Biol Sci Med Sci 2017;72(11):1474–82.

Chapter 6

Neurodevelopment and the gut microbiome

Kaitlyn Oliphant and Jing Lu
Department of Pediatrics, The University of Chicago, Chicago, IL, United States

Introduction

A primer on neurodevelopment

The development of the brain, the complex control center of our coordinated bodily functions and behavior, begins in the womb 3 weeks postconception and continues until at least 20–25 years of age, although debatably ongoing throughout the entire lifespan [1]. The neuron is the main cell type in the brain, and by term birth their creation from neuronal precursor cells, termed neurogenesis, is largely complete [2]. Neurons are responsible for relaying the electrochemical signals necessary to centrally process the world around us in order to make the required informed responses. Indeed, changes in the connectivity of these neurons are paramount to neurodevelopment after birth. The extending axons are directed to the recipient dendrites through sampling the environmental chemical signals; certain biochemicals act as attractants, termed neurotrophic factors, that support the life of the neuron, whereas others are repulsive [1]. Synapses continue to form as a net positive until a maximum is reached at approximately 4 years of age, after which such connections are thought to be gradually pruned back to increase efficiency [3,4]. Different regions and subnetworks of the brain are specialized for particular functions, and each has a specific peak time of development. The hippocampus, responsible for mediating recognition and spatial memory, peaks in growth between 32 weeks gestation and 18 months postnatally [5,6]. The onset of high growth in connectivity for the prefrontal cortex, an area of higher cognition for processing, for example, attention and multitasking behaviors, is in the first 6 months postnatally. Particularly, the monoamine neurotransmitter systems that includes histaminergic, catecholaminergic (dopamine, epinephrine and norepinephrine), and serotonergic signaling to mediate reward, affect and mood, begin to rapidly develop prenatally until 3

The Developing Microbiome. DOI: https://doi.org/10.1016/B978-0-12-820602-7.00006-4

years of age [5,7]. Although the maturation of any one subnetwork is imperative for these behavioral outcomes, perhaps of greater importance is maintaining a coordinated neurodevelopmental trajectory, so that the connections between different subnetworks and brain areas properly form to mediate the more complex cognitive processes [8].

Unlike neurogenesis, the proliferation and maturation of the other brain cells, astrocytes, oligodendrocytes and microglia, is far from complete at birth [9,10]. Oligodendrocytes express neurotrophic factors and an insulating protein called myelin to both promote neuronal survival and efficiency of signal transduction [11–13]. As myelin is white in appearance, the regions of which many myelinated axons extend are termed white matter, whereas the regions of which many of the recipient neuronal cell heads locate are termed gray matter. As white matter forms, the brain increases in volume; the myelination process has the greatest activity during the first 2 years of life, and the brain reaches 90% of adult size by 5 years of age [6,14]. Defects in the formation of white matter as visualized by magnetic resonance imaging are the most frequently observed association with neurodevelopmental delay [15,16]. Astrocytes play a critical role in the integrity of the selectively permeable blood–brain barrier (BBB) but are additionally multifaceted in their ability to participate in inflammatory responses, provide metabolic and mechanical support to neurons, and modulate the formation of synapses [10,17]. Astrocytes begin in a structurally and functionally immature state at birth. In mice, astrocytes continue to generate, divide and expand until 2 weeks postnatally [18,19]. At this stage, process outgrowth and elaboration activity increase until a mature morphology can be observed from 3 to 4 weeks postnatally, which interestingly occurs in parallel with axon and dendrite growth, synaptogenesis and synapse maturation of neurons [20,21]. Put differently, the signals immature astrocytes release have been found to stimulate neuronal growth and synaptogenesis, whereas mature astrocytes promote the maturation of synapses. In humans, the study of astrocyte maturation is difficult due to sample limitations and the complexity of astrocyte morphologies. However, Zhang et al. have demonstrated that purified astrocytes cease to actively proliferate at about 6 months to 1 year postnatally [22]. Further, Sloan et al. cultured astrocyte lineage cells in three-dimensional human cerebral cortical spheroids for 20 months, and they were able to observe a mature astrocyte morphology at day 350, with branching increasing in complexity until the study end period [23]. Immature astrocytes additionally upregulate tight junction protein expression in the BBB, whereas after maturation, these astrocytes connect their end feet to the BBB for supporting capillary endothelial cells and reinforcing tight junctions [24]. It is important to note that the BBB is already functional by embryonic week 14 in humans, but changes in structure and specific substrate permeability continue to occur until at least 20 years of age [24–26]. The final discussed cell type, actually derived from the bone marrow progenitor population and delivered through the BBB, are the microglia [27]. Microglia are important for immune surveillance, safe removal of waste material after apoptosis and also synaptic pruning, as they phagocytose weak synapses [10,27]. Microglia begin in a "premicroglia" state at birth, and gradually transition into an "adult microglia" state until 2 weeks postnatally in mice, as defined by differences in surface markers, morphology and gene expression [28,29]. In humans, this process is complete by birth at 35 weeks gestation based upon morphological evidence [30]. Microglia numbers have also peaked at this stage, and the cell populations will then die back or migrate until the adult signature is observed by 4 weeks postnatally in mice [31]. More gene expression and cell quantification studies are needed to determine when this process occurs in humans [32]. Immature microglia have a greater role in developmental functions, including synaptic remodeling and regulation of the number of neurons, whereas mature microglia are more involved in immune surveillance [29,33]. However, the extent and refinement of activity continues until the stable adult population has been reached; as an example, microglia continue to provide support for myelination through secretion of insulin-like growth factor until postnatal day 22 in mice [31,34,35].

The earliest models of neurodevelopment assumed the formation and disassembly of synaptic connections to be an entirely ordered succession of events programmed by genetics. However, landmark studies by Hubel and colleagues demonstrated that interactions with the environment during a specific time frame are critical [36]. In these studies, it was found that the entire neuronal architecture of an eye of a macaque monkey atrophies, with accompanying vision loss, when sutured shut, but only if the surgery was conducted within the first 6 months of life. Since then, multiple studies have confirmed this phenomenon. Specifically, human children that had been institutionalized in impoverished countries since infancy had on average lower IQs, higher incidences of social and emotional disorders, impaired visually mediated memory and learning, and poorer attention and inhibitory control [37,38]. These children experienced an accumulation of neglected factors, including inadequate nutrition, medical care, socialization, stimulation and comfort from stress. If placed in high-quality foster care before 2 years of age, these children could experience an increase in IQ later in life [39]. Neuronal plasticity is thought to be essential for the organism to adapt to the environment in which it lives, although it carries the caveat that without receiving specific environmental signals at the right time, the organism will fail to reach a maximum potential necessary to thrive or at the extreme survive. As the environment is modifiable, much research interest is focused on which environmental factors are important and when they need to be received, in order to promote neurodevelopment and thus prevent disability. In particular, the gut microbiome presents a promising source for such environmental signals.

The gut–brain axis

The connection between the gut microbiome and the brain, termed the gut–brain axis, has long been recognized but started to gain significantly more traction after experiments in mouse models that were depleted of their microbiota from birth, referred to as germ-free (GF) mice. GF mice exhibit several abnormalities in their brain structure and behavior when compared to wild-type mice of the same strain under identical conditions. Such defects include changes in social and anxiety-like behavior, motor activity and memory [40–44]. These behavioral alterations are accompanied by increased BBB permeability, an altered profile of neurotransmitters, decreased volumes and directionally dependent connectivity in both white and gray matters, reduced myelination, and evidence of neuroinflammation [43,45–48].

In humans, several neurodevelopmental disorders have associated alterations in the structure and function of the gut microbiome, such as autism spectrum disorder (ASD), attention deficit/hyperactivity disorder (ADHD) and schizophrenia [49–52]. In addition, environmental factors that influence gut microbiome succession, including nutrition, geographical location/urban or rural environment, exposure to siblings/pets, and stress, have also been associated with neurodevelopmental outcomes [53–58]. Research has also shown that antibiotics administration in early life is associated with lower cognitive, executive function and verbal comprehension capabilities, increased impulsivity, hyperactivity and incidence of ADHD, and elevated risk of emotional problems, including anxiety and depression [59,60]. Further, studies in which the fecal microbiota of human children is transfaunated into GF mice have yielded neurological phenotypical differences, with examples including cases of preterm infant growth failure and ASD [46,61,62]. Thus, the gut microbiome can act as a biological mediator of environmental factors on the brain.

Plan for the chapter

The existence of the gut–brain axis yields much promise, as the gut microbiome is a readily mutable ecosystem during the early stages of life. Indirect interventions, such as nutritional (e.g., prebiotics) or medical (e.g., antibiotics prescription), or the more direct application of probiotics, could alter the trajectory of gut microbiome succession and, in turn, improve neurodevelopment. However, in order to properly direct such interventions, it is important to first understand the gut–brain axis in the context of development. The plan for this chapter is thus to discuss our current knowledge of the mechanisms by which the gut microbiome can affect neurodevelopment. The influence of the gut microbiota on nutrition will first be examined. The next sections are subdivided by the avenues of communication between the gut and the brain, humoral via the circulatory system, and neural through central nervous system (CNS) connections. Both the direct production of microbial substances and more indirect mechanisms of microbial interaction with intestinal epithelial cells (IECs) will be described. Then, microbial activation of the immune system, including the induction of neuroinflammation, will be discussed in terms of neurodevelopment. Finally, this chapter will conclude with a brief summary and the impact of altered neurodevelopmental trajectories on functioning in adult life.

Microbiome-host interactions impacting neurodevelopment

Nutrition

As the brain has a high caloric demand (20% of total consumed), it is not surprising that malnutrition is one of the most impactful causes of neurological impairment [5,63]. However, several factors can influence the extent and type of impairment that results from nutritional deficiencies. Arguably, the most important is time. The metabolic requirements of the brain are even more voracious during development, for example, it accounts for approximately 50% of the basal metabolic rate of the body at age 10 [64]. Glucose, the energy source for the brain, reaches a maximum uptake when synaptic density peaks (roughly 4 years of age); thus, malnutrition has a disproportionality greater influence on neurodevelopment during these time periods [3,4,65]. In addition, distinct neuronal networks are formed at specific ages. For example, an earlier environmental insult would more likely affect the hippocampus, considering that its peak growth is during the first 18 months of life, which would impair brain circuits that are dependent upon balanced signaling between the hippocampus and cortex, such as the ventral tegmental area loop implicated in schizophrenia [5,6,66]. Other factors shaping the effect of malnutrition on neurodevelopment include the severity, duration, type of nutrient deficiency and notably, the environment [67]. Thus, a simple linear relationship between calorie restriction and neuronal degeneration does not accurately describe how malnutrition impacts neurodevelopment, and indeed the gut microbiota is known to play a crucial mediating role.

One of the most important functions of the gut microbiota is to degrade otherwise indigestible food material into nutrients usable by the host, thus modulating the calories yielded by the three macronutrients, carbohydrates, proteins and fat. The major end products of microbial carbohydrate fermentation pathways are short-chain fatty acids (SCFAs), of which acetate, propionate and butyrate are most abundant [68]. IECs can directly utilize SCFAs as a source of fuel in a preferential order, first butyrate then propionate and acetate. In adults, this metabolic process generates 10% of the required caloric content [69]. In addition, SCFAs that escape catabolism by IECs can enter hepatic circulation to affect host metabolism globally. Specifically, acetate and butyrate are lipogenic, whereas propionate is gluconeogenic [68]. As the brain is dependent upon glycolysis and aerobic glycolysis for ATP generation, the gut microbiota can influence the efficiency of these metabolic pathways by modulating the amount of free glucose, nicotinamide–adenine dinucleotide cofactors, and lactate available in circulation, through both their own metabolic processes and host metabolic control [64]. Such effects are not only significant for undernutrition, as the gut microbiota is also necessary for the development of hyperglycemia and insulin resistance in cases of overnutrition, which causes anxiety- and depressive-like behaviors in mice [70]. Indeed, a brain-specific knockout of the insulin receptor in mice led to augmented dopamine turnover, and mood has also been reported to improve in obese humans after insulin injection [71,72]. The degradation of proteins by the gut microbiota can alter the pool of amino acids available to the host, and Navarro et al. observed that this phenomenon was related to the capability of certain individuals to maintain nitrogen balance and weight during periods of protein deficiency [73]. SCFAs are also the most abundant end products of microbial amino acid fermentation; however, a plethora of other intermediate by-products are additionally produced, including the phenolic, indolic and neurotransmitter compounds discussed later [74]. The small intestinal microbiota, in particular, has been found to play a role in fat digestion and absorption, as GF mice have increased fecal lipid excretion, and microbial bile acid biotransformation has been proposed as a potential mechanism for this activity [75,76]. Microbes in the gastrointestinal (GI) tract are also capable of modulating other nutrients important for brain development, including the bioavailability of choline and the synthesis of B vitamins [5,77,78]. Further, microbes can promote the excretion of dietary neurotoxins, such as heavy metals through the production of siderophores [79].

Clearly, the gut microbiota affects how nutrients are absorbed and utilized by the host. However, when examining the directionality of the relationship, it is also important to note that diet is a dominant modulator of gut microbiota composition, as the availability of substrates would naturally impact the ability of a microbial species to survive and thrive in a host's gut [80]. Indeed, malnourished infants and children have a delayed pattern of gut microbiome succession when compared to relevant controls [81,82]. Therefore, a feasible route of intervention that exploits the diet-gut microbiota-development axis was demonstrated in a study by Gehrig et al. [83]. Their group was able to derive a set of microbes that promoted weight gain in GF mice when compared to appropriate controls, from modeling the gut microbiome succession of children with good and poor growth. They then determined which microbiota-directed foods selectively enriched the growth of these microbes in mouse models, and verified that these foods were able to promote their colonization of the intestinal environment of Bangladeshi children with moderate acute malnutrition. In turn, these children experienced improved growth, immunity and bone formation, in addition to elevated concentrations of neurodevelopmental markers in their serum. Taken together, the gut microbiota can act as a mediator of nutrition that impacts neurodevelopment.

Humoral communication mechanisms

In addition, recent evidence suggests a more complex relationship between microbial metabolism and neurodevelopment. Certain metabolites produced in excess by the gut microbiota are able to pass through the epithelial lining of the intestine and enter the bloodstream, which contributes to roughly 10% of the total chemical species found in human circulation [84]. In turn, the humoral landscape of the brain is altered, as some of these metabolites can interact with the endothelial cells lining the capillaries of the brain to affect permeability of substrates, or even cross the BBB to interact with brain cells directly. SCFAs, phenolic and indolic compounds, bile acids, and certain microbial surface structures have been found to have this capability [85–90].

SCFAs are able to regulate gene expression in the brain through inhibition of histone deacetylases, which affects epigenetic programming, and activation of G protein–coupled receptors [87]. As a histone deacetylase inhibitor, administration of sodium butyrate alone has been shown to improve neurological outcomes in animal models, including (1) reduced development of absence epilepsy and behavioral comorbidities (depressive-like behavior and cognitive performance) in WAG/Rij rats when therapy was started prior to absence seizure onset [91], in addition to preventing temporal lobe epilepsy in hippocampus kindling mouse models [92], (2) corrected memory and learning deficits in maternally deprived mice, a model for early life stress [93], and (3) enhanced neurogenesis and increased granular cell layer volume in the hippocampus of piglets [94]. Indeed, Danilova et al. demonstrated that acetylation of histone H3 is critical for long-term memory formation in a snail model, in part through administering the histone deacetylase inhibitor, sodium butyrate, to correct for learning defects in juvenile snails [95]. They additionally determined that histone H3 acetylation is regulated by extracellular signal-regulated kinases, and Shah et al. observed that butyrate was able to activate the extracellular signal-regulated/mitogen-activated protein kinase (MAPK) signaling pathways in the PC12 cell line (pheochromocytoma of the rat adrenal medulla), which also led to increased expression of tyrosine hydroxylase to affect dopaminergic signaling [96]. The effects of butyrate can extend beyond neurons as well, as Huang et al. found that histone H3 acetylation was correlated with the differentiation and maturation of oligodendrocytes in a neonatal rat model of white matter injury, in which sodium butyrate administration rectified these defects [97]. Intriguingly, a time dependency for the histone deacetylase inhibitory activity of butyrate has been noted, which varied from a marked stimulation in the cerebral cortex of young rats to no discernable effect by old age [98].

Taken together, butyrate, as a histone deacetylase inhibitor, appears to support myelination by promoting the maturation of oligodendrocytes, enhance learning and long-term memory retention through epigenetic modifications, and modulate mood through regulation of dopaminergic signaling. Propionate was later found to have a similar activity on the dopaminergic axis, in addition to being the only SCFA able to induce tryptophan 5-hydroxylase expression in the PC12 cell line, thus altering serotonergic signaling [99]. A dose-dependency was additionally uncovered in this study, as high concentrations of butyrate and propionate actually inhibited tyrosine hydroxylase. The latter observation suggests that an optimal range of SCFAs must be present, as interestingly, direct injection of propionate into the brain of rats impairs social behavior [100]. In addition, the human brain endothelium expresses receptor FFAR3, which has affinity for propionate and butyrate, and activation by these substrates was able to reduce BBB permeability after challenge with an inflammatory trigger in vitro [101]. Administration of sodium butyrate to GF mice also increased the expression of the tight junction protein occludin in the frontal cortex and hippocampus [45]. Finally, an additional mechanism discovered by which SCFAs can exert their effects is through direct metabolism by brain cells. Acetate can be utilized by astrocytes in the hypothalamus, which supports the glutamate–glutamine transcellular cycle and thus GABAergic signaling to ultimately suppress appetite [102]. Therefore, SCFAs appear to have a complex and multifaceted impact on neurodevelopment of the host, warranting future investigation.

The metabolism of aromatic amino acids by the gut microbiota yields a plethora of phenolic and indolic compounds, of which phenyl sulfate, pyrocatechol sulfate, 3-(3-sulfoxyphenyl)propanoic acid and indoxyl sulfate were found to be associated with fear extinction learning in a study by Chu et al. [88]. They observed that GF mice had impaired learning-related remodeling of postsynaptic dendritic spines in the medial prefrontal cortex, which resulted in reduced activity. This defect was related to changes in gene expression of microglia. Microglia play important roles in developing social and cognitive neurocircuitry, as their phagocytotic activity can prune the formation of synapses [103]. Further, the gut microbiota promotes the maturation of microglia, as GF mice exhibit a skewed profile toward the prenatal type [9]. Intriguingly, Chu et al. confirmed a time dependency for this microglial effect, as only GF mice that were provided access to natural mouse microflora prior to weaning had a rescued ability to learn [88]. Both SCFAs and afferent vagal signaling have been proposed as mediators of microglial functioning [33]. However, Chu et al. demonstrated that phenolic and indolic metabolites in the cerebrospinal fluid differentiated GF mice from controls [88]. Further, a phenolic metabolite, 3-(3-hydroxyphenyl)-3-hydroxypropionic acid (HPHPA), a derivative of 3-(3-sulfoxyphenyl)propanoic acid, has been detected in increased amounts in urine from autistic and schizophrenic patients when compared to relevant controls [104,105]. Clayton was able to propose an intestinal microbial link for the production of HPHPA, based on a previously observed metabolic defect in laboratory rats termed "chlorogenic acid" phenotype [106]. These rats resulted from housing under microbiologically restricted environments, and were suggested to exhibit altered GI catabolism of aromatic amino acids from an overabundance of *Clostridium* and *Peptostreptococcus,* which could be converted back to the normal phenotype after being cohoused with rats of that phenotype. In addition, Hsiao et al. demonstrated that pups born from the maternal immune activation model of autism exhibited an abnormal gut microbiota composition, increased intestinal permeability, irregular behavior, and an altered serum metabolite profile, which could be prevented by treatment with a strain of *Bacteroides fragilis* at weaning [107]. The phenolic metabolite 4-ethylphenylsulfate exhibited the greatest concentration change after treatment in the serum of these mice, and direct injection of 4-ethylphenylsulfate caused anxiety-like behavior when administered to mouse pups from 3 to 6 weeks of age. Interestingly, microbial phenolic metabolites derived from dietary polyphenols reduced lipopolysaccharide (LPS)-induced production of nitric oxide and cytokines by microglia in vitro, demonstrating an antiinflammatory effect [108]. This activity is the proposed mechanism for the observed therapeutic effects of dietary polyphenol intake in patients with depression [109]. Also, microbial tryptophan metabolites in general, including indole, indole-3-acetate and tryptamine, can interact with the aryl hydrocarbon receptor (AHR), and deletion of this receptor in mice leads to impaired hippocampal-dependent contextual fear memory, with decreased neurogenesis, neuronal migration and differentiation of neuronal cell types [110,111]. The AHR additionally plays a role in immune surveillance and the dampening of the immune response [112,113]. Indeed, Rothhammer et al. confirmed that microbial metabolites can act as agonists for the AHR expressed by microglia, which suppressed neuroinflammation in mouse models through downregulating expression of cytokines that control astrocyte activity [114]. Thus phenolic and indolic compounds can influence the behavior of microglia, and since dysfunction in microglial-mediated synaptic pruning has been associated with both autism and schizophrenia, it is worth investigating how these compounds interact with microglia during development [115,116].

Other microbial by-products have just begun to be studied for their effects on the brain and present an exciting avenue for future research. The biotransformation of bile acids by the gut microbiota produces the secondary bile acids, which have been found to affect BBB permeability and dopaminergic signaling via the brain bile acid receptor Takeda G protein−coupled receptor (TGR5) [117−120]. Developmentally, dopamine impacts the frontal cortex and striatal neurocircuitry through regulation of the proliferation, migration and differentiation of the associated subpopulations of neurons, ultimately affecting reward-related behavior, cognition and mobility [121].

Peptidoglycan, present in all bacterial cell walls but more abundant in Gram-positive bacteria, can cross the BBB under normal conditions and interacts with peptidoglycan-recognition proteins to influence social behavior [85]. LPS, present on the cell surface of Gram-negative bacteria, is a potent activator of the inflammatory cascade that can also cross the BBB [90]. Further microbially derived substances that impact brain development will likely be discovered as research progresses in the field.

Neural communication mechanisms

The human intestinal tract is densely innervated. Most of the present neurons comprise the enteral nervous system (ENS) that is in and of itself an independent system that controls motility and intestinal absorption [122]. However, the ENS is connected to the CNS, enabling a route of central communication [123]. The CNS can also directly interact with stimulatory and inhibitory neuroactive substances in the gut via the vagus nerve [122–124]. Afferent fibers of the ENS and CNS do not cross the epithelial barrier, so microbial cells cannot interact with them directly [125]. However, neurotransmitters produced by the gut microbiota can pass through the intestinal barrier, and microbial interaction with enteroendocrine cells and the immune system impacts vagal and enteric neural transmission [122–124]. Such signals have been found to be involved in the regulation of hunger, as well as inflammation and "sickness" behaviors, including depression, anxiety, anorexia, fatigue and pain sensitization [123]. Indeed, a vagotomy can abolish certain gut microbiota–mediated effects on host behavior [123,124]. In this section the effects of two major systems of neuronal signaling, glutamic and serotonergic, are to be discussed, with others mentioned as areas of future work.

The glutamate *N*-methyl-D-aspartate (NMDA) receptors in the brain play a role in promoting neuronal outgrowth, survival and synaptogenesis during development [126]. NMDA receptor hypoactivation has thus been implicated in learning disabilities and schizophrenia, and the timing of the insult is especially important in this case, which determines the extent and region of the brain that is affected [127,128]. Particularly, administration of NMDA receptor antagonists during the first 2 weeks of life in rats (i.e., the brain growth spurt period, preweaned) triggers apoptotic neurodegeneration, resulting in spatial learning and working memory defects, hyperlocomotion, and altered sensorimotor gating [127]. However, NMDA receptor antagonists, chiefly gamma-aminobutyric acid (GABA), are also known to prevent stress, anxiety and depression-like behavior in animal models [129]. Similarly, NMDA hyperactivation can induce neurotoxicity during development, and administration of NMDA receptor agonists during the first 2 weeks of life in rats demonstrated an especially concentrated cell death in the hippocampus, with putative increases in repetitive behaviors [129,130]. It therefore appears that a balance between glutamate and GABA is required. GF mice exhibit downregulation of the NMDA receptor coupled with anxiolytic behavior, thus demonstrating a role for the gut microbiota in regulating glutamic and GABAergic signaling [131]. This activity may stem from the ability of the gut microbiota to directly modulate glutamate and GABA pools. Glutamate may be released from intestinal protein degradation, and several species of bacteria are able to ferment glutamate and produce GABA, such as the early colonizers *Bifidobacterium* and *Lactobacillus* [73,74,132]. The vagus nerve may be a critical route of communication in this case, as GABAergic signaling induced by a strain of *Lactobacillus rhamnosus* that ameliorated stress, anxiety and depression in mice was completely abrogated after a vagotomy [40]. However, another indirect mechanism by which the gut microbiota may exert its effects on NMDA receptor transmission is through control of brain-derived neurotrophic factor (BDNF) levels. NMDA receptors promote the expression of BDNF, and, in turn, BDNF can modify the subunit compositions and expression of NMDA receptors [133]. BDNF levels are altered in schizophrenic and autistic patients, and are associated with changes in anxiety, depressive-like behaviors, memory and learning in mice [134–137]. BDNF thus unsurprisingly also has a known role in the development of neuronal circuits through supporting neuronal survival and differentiation, as well as regulation of synapses and plasticity [133]. GF mice have lower concentrations of BDNF in their hippocampus, prefrontal cortex and amygdala, but higher concentrations specifically in their dentate gyrus [40,131]. In addition, colonization of mice with *Bifidobacterium* has the ability to restore BDNF levels [48,138]. SCFAs in particular have a demonstrated capability for modulating BDNF expression, as Wei et al. reported decreased methylation of the *bdnf* gene in mouse brains after oral sodium butyrate administration [139]. Therefore the connection between GABA, BDNF, the gut microbiota, and proper activation of NMDA receptors during development warrants further investigation.

Serotonin, a neurotransmitter involved in the regulation of food intake, bone remodeling and anxiety-/stress-related behaviors, is additionally known to play a key role in neurodevelopment [140]. It can regulate neurogenesis, neuronal growth and differentiation, and synaptogenesis [141]. Particularly, serotonin can modulate the secretion of Reelin, a protein that directs neuronal migration, and this can shape neuronal microcircuitry [142]. Serotonin dysregulation thus has a number of neurodevelopmental consequences, including anxiety, depression, social deficits and autism, and motor impairments [143]. Timing of the disturbance is a key element in understanding the neurological outcomes of serotonin dysregulation, for example, motor impairments are restricted to the early postnatal period when somatosensory barrel cortex formation takes place [143–146]. This concept is further illustrated by selective serotonin reuptake inhibitor drugs that are used to treat depression, anxiety and antisocial behavior in adults, and yet paradoxically, have been found to cause these same behavioral disorders later in life when administered pre- or postnatally [143]. Metabolites produced by the gut microbiota, including butyrate, propionate, secondary bile acids and tryptamine, can stimulate the production of serotonin by enteroendocrine cells and its subsequent serum bioavailability [140]. Further, the branched-chain fatty acid products of branched-chain amino acid fermentation, isovalerate and isobutyrate, were found to stimulate enteroendocrine cells via olfactory receptor 558 [147]. GF mice thus unsurprisingly have elevated adenocorticotropic hormone and corticosterone in response to stress and reduced serotonin levels [48,148]. A heightened stress response induced by maternal separation in mice and rats was found to delay neurodevelopment particularly through downregulation of proteins involved in the synaptosome [149,150]. Indeed, GF mice also exhibit downregulation of synaptophysin and PSD-65 [151]. Sudo et al. found that monocolonization of GF mice with a strain of *Bifidobacterium infantis* was able to reverse the abnormal hypothalamic-pituitary-adrenal (HPA) axis response, whereas monocolonization with a strain of *Escherichia coli* did not [48]. The authors proposed mechanism for the microbial-HPA axis interaction involved the induced expression of the cytokine interleukin-1β (IL-1β) by the *E. coli* strain, and BDNF plus the vagal transcriptional regulator c-Fos by the *B. infantis* strain. Multiple cytokine families, including IL-1β, are capable of elevating the production of hormones in the HPA axis [152]. Meanwhile, the activity of the *B. infantis* strain was not found to be immune mediated, since administration of anti-interleukin-6 (IL-6) antibody did not prevent restoration of HPA axis normalcy. The full mechanism is thus not entirely clear, and the subject of future work.

Although less well described, other neurotransmitters and hormones can be modulated by the gut microbiota and should be examined in future work. For example, the neurotransmitter histamine can be produced from fermentation of the amino acid histidine by *Proteobacteria* and *Lactobacillus* [153]. The histaminergic system is involved in modulating wakefulness, motor control, pain perception, learning and memory [154]. In addition, histamine has been found to regulate proliferation and differentiation of neural stem cells into GABA- and neuronal nuclear protein-positive neurons [155]. Histamine is known for its role in exacerbating inflammation in allergenic responses, and in a mouse model of preterm infant white matter injury, administration of a histamine inhibitor attenuated myelination defects, and macro- and microgliosis [156]. However, it is unclear whether bacterially produced histamine in the intestine is proinflammatory, as it has contradictorily been shown to inhibit production of key cytokines both in vitro and in vivo, or if it can indeed influence neurodevelopmental trajectories [157,158]. In addition, the anorexigenic hormones, including glucagon-like peptide 1, peptide YY and leptin, can be produced by enteroendocrine cells after activation from microbially derived SCFAs and secondary bile acids [87]. Although the regulation of appetite is a modulator of nutrition, as, for example, gut hormones are correlated with feeding intolerance in preterm infants, it remains to be elucidated as to whether or not there are farther reaching impacts of these hormones on host development [159]. As research continues in the field, further mechanisms by which the gut microbiota can regulate neurological substances will be uncovered.

The immune system and neuroinflammation

The gut microbiota plays critical roles in the maturation of the immune system, since microbial cell surface structures can act as microbial-associated molecular patterns (MAMPs) that interact with pattern recognition receptors to communicate with the host immune system, and microbial metabolites can have both pro- or antiinflammatory effects [160,161]. Thus, the presence of the gut microbiome changes the profile of serum cytokines, which further alters the humoral factors available to the brain [162]. Cytokines have been found to act as messengers for cells in the brain and are critical to neurodevelopment [163]. Under normal circumstances, the gut microbiota is able to stimulate the correct levels of cytokine production required for optimal brain development, which are known to be altered in GF mice [164,165]. However, it is also imperative to note that, from infancy and until 2 years of age, the immature immune system is hyporesponsive to low levels of stimulus [166]. This diminished sensitivity in surveillance can allow for the proliferation of opportunistic pathogens, and once the amount of stimulus reaches a certain threshold, the mounted immune response can be as effective as an adult. Therefore, this age group is paradoxically at increased risk for aggressive immune responses and even sepsis [166,167]. Such responses can breach the integrity of the intestinal barrier, leading to increased circulation of MAMPs, including LPS, and proinflammatory cytokines, such as IL-1β and IL-6 [168–171]. These factors can render the brain susceptible to mounting an inflammatory response, termed neuroinflammation. Evidence of this phenomenon comes from the observed associations between early infections (not only including the obvious example of meningitis) or necrotizing enterocolitis (an intestinal inflammatory disease occurring in premature infants) and poor neurological outcomes, including cerebral palsy, and other psychomotor, visual and cognitive impairments [172–174]. However, it should be noted that even in the absence of disease, inflammatory markers are associated with impaired neurodevelopment, including motor, language and nonverbal intelligence scores [171,175,176]. Malnutrition and systemic inflammation are also intertwined, as GI inflammation can prevent the absorption of key nutrients; for example, infection elevates levels of hepcidin resulting in reduced iron absorption [175,177]. In cases of low-level systemic inflammation, it is not one pathogen alone that is contributing to chronic inflammation, but rather, configurations of the intestinal microbiota that differ from the norm. The consequences are disproportionate interactions with toll-like receptors (TLRs) and production of metabolites with pro- and antiinflammatory effects. The most frequently observed alteration for pediatric inflammatory bowel diseases is an increase in the proportion of *Proteobacteria*, which activate TLR4 via LPS present on their cell surface [178,179].

Neuroinflammation, like inflammation in other parts of the body, is a conglomerate of innate and adaptive responses that relies upon circulating immune system components, in addition to the resident astrocytes and microglia populations [180]. Thus, the BBB must yield to allow for the penetration of these necessary circulating components during events of neuroinflammation. Intriguingly, even though young animals were found to be resilient against increases in BBB permeability induced by LPS challenge, unlike adults, Stolp et al. observed that early immune system activation primed rats for increased BBB permeability later in life without an additional stimulus, which resulted in behavioral changes [181,182]. Indeed, several neurodevelopmental disorders have a reported increase in BBB permeability, including autism, schizophrenia, and mood disorders [183]. The juvenile BBB is especially permeable to plasma proteins, however, specifically the acute phase reactants [184]. As microglia, astrocytes and oligodendrocyte precursor cells contain receptors for IL-1β and interferon-γ, excessive amounts of proinflammatory cytokines can have a great impact on brain functioning [185]. Microglia and astrocytes are also able to amplify the production of proinflammatory cytokines in response [186]. Although the absence of gut microbiota has not been found to have an effect on the mature astrocyte population in adult GF mice, neuroinflammation induces the expression of tissue growth factor-β (TGF-β) to regulate immune and injury-related responses [9,187]. The TGF-β superfamily is referred to as the brain morphogenic proteins that are involved in regulating neuronal induction [163]. TGF-β inhibits self-renewal of neural progenitor cells stimulated by Wnt and fibroblast growth factor signaling, and promotes subtype differentiation in a gradient-specific manner [188]. TGF-β is also able to inhibit the production of oligodendrocyte precursors and induce the differentiation of astrocytes specifically in relation to BBB maturation [189,190]. Thus the consequence of early and excess TGF-β concentrations could be an irreversible switch in the role of astrocytes that occurs prematurely [190]. As immature astrocytes promote the expression of tight junction proteins, it is most interesting that Stolp et al. discovered decreased levels of the BBB tight junction protein claudin-5 after early immune activation [24,182]. Evidence of neuroinflammation and increased TGF-β has also been observed in the brains of patients with autism [191]. Further, overexpression of TGF-β in mice induced antisocial and repetitive behaviors, but only when induced in preweaned mice and not in adults [192]. As immature astrocytes promote BBB integrity and the formation of neural networks, their role in these phenotypic changes should be explored [21,24]. Further, the timing and induction of normal concentrations of TGF-β by the gut microbiota during development should be examined. For example, *Clostridia* have been found to be potent activators of TGF-β production, which has been linked to TLR2 signaling in dendritic cells, inhibition of metalloproteases and production of SCFAs, and thus these species may represent particularly important colonizers [193]. In the extreme, TGF-β knockout mice develop congenital defects (including craniofacial) and may perish early, both of which vary upon the TGF-β subtype and genetic background of the mouse [194].

Although a delayed maturation and the developmental roles of microglia have previously been discussed, both microglia recruitment and immunological activity are necessarily enhanced during neuroinflammation [180]. Part of this microglial activity is the production of reactive oxygen species, and autopsies of infants with white matter injury particularly exhibited oxidative damage of progenitor premyelinating oligodendrocytes [195,196]. An accumulated damage caused by neuroinflammation is of higher concern, since the brain has a limited regenerative capacity compared to other organs [197]. Microglia are additionally potent producers of IL-6 during early phases of CNS inflammation [198]. The IL-6 superfamily is referred to as the neuropoietic cytokines that have pleiotrophic effects on neurons, astrocytes, and oligodendrocytes, including the promotion of proliferation, differentiation/fate switching, survival and apoptosis [163]. A particularly important role for IL-6 is the gliogenic switch for neurogenic progenitors, as mice for which its receptor in the brain, gp130, is knocked-out have impaired cortical astrocyte production [199]. IL-6 is also able to increase the numbers of oligodendrocytes [189]. Interestingly IL-6, both pre- and postnatally, has been directly implicated in the neurological abnormalities resulting from early life immune activation, including autism-like behaviors [200,201]. Fortunately, certain microbial metabolites are able to deescalate such harmful activity. In addition to the previously discussed phenolic and indolic metabolites, microglia express the bile acid receptor TGR5, which upon binding reduce neuroinflammation both in vitro and in vivo [202]. Thus future work should focus on microbial mechanisms of regulating uncontrolled inflammatory responses, especially via microglia. However, it should be noted that a transient activation of IL-6 in early life is an apparently necessary stage in development, as one of the most prominent early colonizers, *Bifidobacterium*, is able to induce IL-6 production by IECs via NF-κB and MAPK signaling in a TLR2-dependent manner [203]. The effects of a delay in the timing of this activation should additionally be examined further. For example, IL-6 knockout mice are resistant to stress and depression-like behaviors, but are increasingly hedonistic and develop mature-onset obesity [204,205].

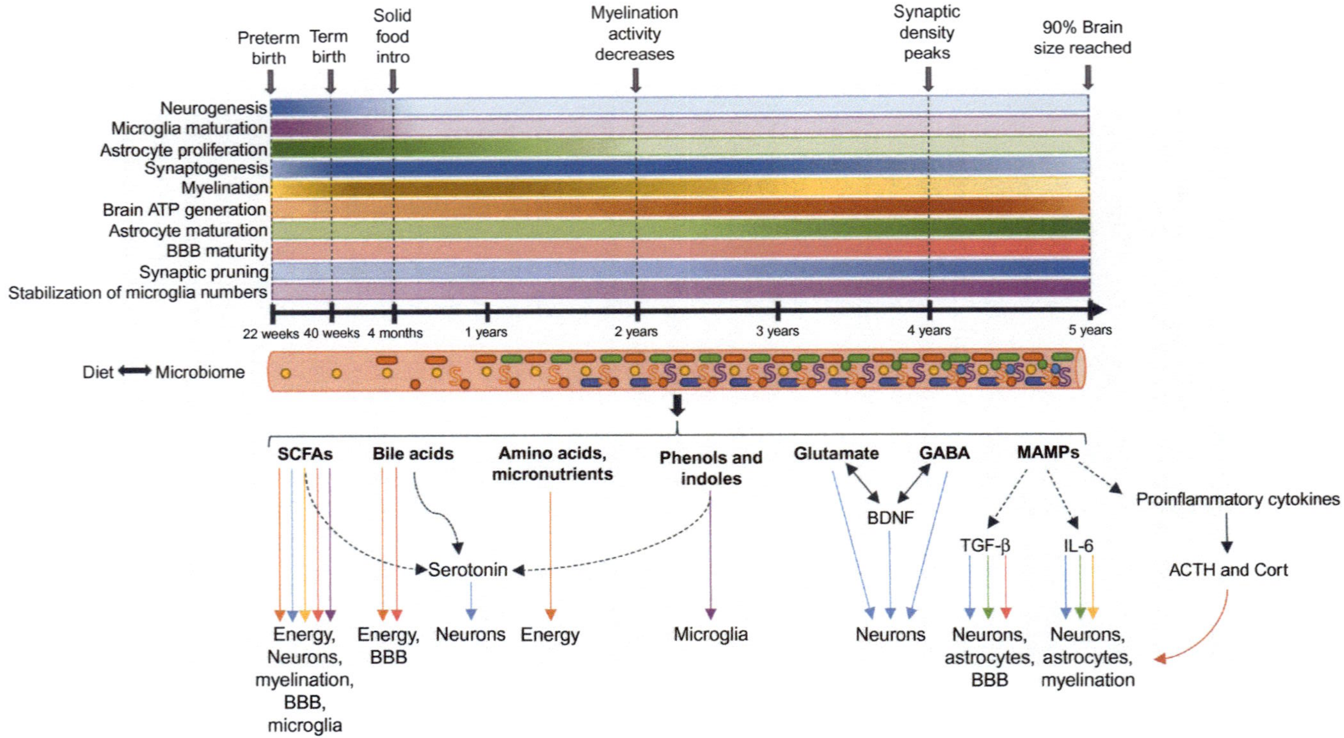

FIGURE 6.1 Summary of the mechanisms by which the gut microbiota influences neurodevelopment. Neuronal processes are indicated in blue, brain energy requirements in orange, myelination activity in yellow, BBB maturity in pink, astrocyte processes in green, and microglia processes in purple. Synaptogenesis and myelination have greater activity early in life, reaching a maxima at around 2–3 years, and decreasing heavily after synaptic density peaks at 4 years of age. Brain ATP generation, which requires uptake of glucose and other nutritional cofactors, is highest at this synaptic density peak. As synapses mature, so do astrocytes that promote this activity, which occurs largely after completion of astrocyte proliferation at 6 months to 1 year of age. In turn, the BBB also matures. Once synaptic density peaks, synaptic pruning increases in relative activity. Microglia mature in parallel with neurogenesis and cell populations stabilize due to die back or migration as synaptic remodeling is completed, in order to switch into their new role from synaptic pruning to immunosurveillance. Meanwhile, the gut microbiota is increasing in diversity from birth and produces metabolites that influence all these processes, as indicated by the color-coordinated arrows. Inflammation and the stress response, however, inhibit neurodevelopment. MAMPs stimulate the immune system, and all metabolites listed here may either promote tolerance or aggravate the response. Proinflammatory cytokines can then induce the release of the hypothalamic-pituitary-adrenal hormones, ACTH and Cort. *ACTH*, Adrenocorticotrophic hormone; *BBB*, blood–brain barrier; *Cort*, cortisol; *MAMPs*, microbial-associated molecular patterns.

Implications for lifetime health

There are multiple mechanisms by which the gut microbiome can influence neurodevelopmental outcomes (Fig. 6.1). The gut microbiota can affect how energy is derived and stored from food, produce substances available to the brain from systemic circulation, interact with host immune and enteroendocrine cells, and alter neuronal signaling. The combined impact results in changes to the evolving neurocircuitry, during a critical time period of a child's life at which neuronal plasticity takes place. Lifelong effects on an individual's sociability, anxiety, capability of learning, cognition and mobility are the consequence. At worst, there is a risk of neurodevelopmental disability, including ASD, ADHD, learning disability, schizophrenia, or cerebral palsy. The CDC has reported that the prevalence of developmental disability is increasing in the United States, reaching 17.8% between 2015 and 2017 [206]. There is a considerable cost burden associated with medical care, behavioral therapy, special education services and loss of productivity for both the parents and individual upon adulthood. Further, developmental disabilities can also present a significant comorbidity burden. As an example, 90% of patients with ASD have reported feeding or GI disorders, and about a third have reported immune dysregulation (autoimmune disease and/or allergy) and metabolic/cardiovascular diseases (obesity, diabetes, hypertension and/or dyslipidemia) [207,208]. Intriguingly, the known functions of the gut microbiota in aiding digestion, regulating host metabolism and modulating the immune system are thought to have an important role in the development and maintenance of GI, inflammatory and metabolic disorders [209–211]. The early gut microbiota, in particular, is responsible for educating the immune system to facilitate tolerance to environmental exposures, and antibiotic use in the first year of life is associated with an increased risk of developing allergy and asthma, inflammatory bowel disease and metabolic syndrome [212–216]. Targeting the developing gut microbiota may thus prove to be a holistic approach to optimize the health and wellness of an individual throughout their entire lifespan.

References

[1] Stiles J, Jernigan TL. The basics of brain development. Neuropsychol Rev 2010;20:327–48.

[2] Snyder JS. Recalibrating the relevance of adult neurogenesis. Trends Neurosci 2019;42:164–78.

[3] Huttenlocher PR, Dabholkar AS. Regional differences in synaptogenesis in human cerebral cortex. J Comp Neurol 1997;387:167–78.

[4] Liu X, Somel M, Tang L, Yan Z, Jiang X, Guo S, et al. Extension of cortical synaptic development distinguishes humans from chimpanzees and macaques. Genome Res 2012;22:611–22.

[5] Cusick SE, Georgieff MK. The role of nutrition in brain development: the golden opportunity of the "first 1000 days. J Pediatr 2016;175:16–21.

[6] Rice D, Barone S. Critical periods of vulnerability for the developing nervous system: evidence from humans and animal models. Environ Health Perspect 2000;108(Suppl. 3):511–33.

[7] Field T, Diego M, Hernandez-Reif M, Figueiredo B, Deeds O, Ascencio A, et al. Prenatal dopamine and neonatal behavior and biochemistry. Infant Behav Dev 2008;31:590–3.

[8] White NM, McDonald RJ. Multiple parallel memory systems in the brain of the rat. Neurobiol Learn Mem 2002;77:125–84.

[9] Erny D, Hrabě de Angelis AL, Jaitin D, Wieghofer P, Staszewski O, David E, et al. Host microbiota constantly control maturation and function of microglia in the CNS. Nat Neurosci 2015;18:965–77.

[10] Jiang X, Nardelli J. Cellular and molecular introduction to brain development. Neurobiol Dis 2016;92:3–17.

[11] Du Y, Dreyfus CF. Oligodendrocytes as providers of growth factors. J Neurosci Res 2002;68:647–54.

[12] Hermoye L, Saint-Martin C, Cosnard G, Lee S-K, Kim J, Nassogne M-C, et al. Pediatric diffusion tensor imaging: normal database and observation of the white matter maturation in early childhood. NeuroImage 2006;29:493–504.

[13] Oppenheim RW. The neurotrophic theory and naturally occurring motoneuron death. Trends Neurosci 1989;12:252–5.

[14] Lenroot RK, Giedd JN. Brain development in children and adolescents: insights from anatomical magnetic resonance imaging. Neurosci Biobehav Rev 2006;30:718–29.

[15] Anderson PJ, Cheong JLY, Thompson DK. The predictive validity of neonatal MRI for neurodevelopmental outcome in very preterm children. Semin Perinatol 2015;39:147–58.

[16] Hinojosa-Rodríguez M, Harmony T, Carrillo-Prado C, Van Horn JD, Irimia A, Torgerson C, et al. Clinical neuroimaging in the preterm infant: diagnosis and prognosis. NeuroImage Clin 2017;16:355–68.

[17] Stolp HB, Liddelow SA, Sá-Pereira I, Dziegielewska KM, Saunders NR. Immune responses at brain barriers and implications for brain development and neurological function in later life. Front Integr Neurosci 2013;7:61.

[18] Ge W-P, Miyawaki A, Gage FH, Jan YN, Jan LY. Local generation of glia is a major astrocyte source in postnatal cortex. Nature 2012;484:376–80.

[19] Tien A-C, Tsai H-H, Molofsky AV, McMahon M, Foo LC, Kaul A, et al. Regulated temporal-spatial astrocyte precursor cell proliferation involves BRAF signalling in mammalian spinal cord. Development 2012;139:2477–87.

[20] Bushong EA, Martone ME, Ellisman MH. Maturation of astrocyte morphology and the establishment of astrocyte domains during postnatal hippocampal development. Int J Dev Neurosci 2004;22:73–86.

[21] Farhy-Tselnicker I, Allen NJ. Astrocytes, neurons, synapses: a tripartite view on cortical circuit development. Neural Dev 2018;13:7.

[22] Zhang Y, Sloan SA, Clarke LE, Caneda C, Plaza CA, Blumenthal PD, et al. Purification and characterization of progenitor and mature human astrocytes reveals transcriptional and functional differences with mouse. Neuron 2016;89:37–53.

[23] Sloan SA, Darmanis S, Huber N, Khan T, Birey F, Caneda C, et al. Human astrocyte maturation captured in 3D cerebral cortical spheroids derived from pluripotent stem cells. Neuron 2017;95:779–790.e6.

[24] Saili KS, Zurlinden TJ, Schwab AJ, Silvin A, Baker NC, Hunter III. ES, et al. Blood-brain barrier development: systems modeling and predictive toxicology. Birth Defects Res 2017;109:1680–710.

[25] Semple BD, Blomgren K, Gimlin K, Ferriero DM, Noble-Haeusslein LJ. Brain development in rodents and humans: identifying benchmarks of maturation and vulnerability to injury across species. Prog Neurobiol 2013;106–107:1–16.
[26] Virgintino D, Errede M, Robertson D, Capobianco C, Girolamo F, Vimercati A, et al. Immunolocalization of tight junction proteins in the adult and developing human brain. Histochem Cell Biol 2004;122:51–9.
[27] Kettenmann H, Kirchhoff F, Verkhratsky A. Microglia: new roles for the synaptic stripper. Neuron 2013;77:10–18.
[28] Bennett ML, Bennett FC, Liddelow SA, Ajami B, Zamanian JL, Fernhoff NB, et al. New tools for studying microglia in the mouse and human CNS. Proc Natl Acad Sci USA 2016;113:E1738–46.
[29] Matcovitch-Natan O, Winter DR, Giladi A, Aguilar SV, Spinrad A, Sarrazin S, et al. Microglia development follows a stepwise program to regulate brain homeostasis. Science 2016;353.
[30] Esiri MM, al Izzi MS, Reading MC. Macrophages, microglial cells, and HLA-DR antigens in fetal and infant brain. J Clin Pathol 1991;44:102–6.
[31] Lenz KM, Nelson LH. Microglia and beyond: innate immune cells as regulators of brain development and behavioral function. Front Immunol 2018;9.
[32] Menassa DA, Gomez-Nicola D. Microglial dynamics during human brain development. Front Immunol 2018;9.
[33] Abdel-Haq R, Schlachetzki JCM, Glass CK, Mazmanian SK. Microbiome-microglia connections via the gut-brain axis. J Exp Med 2019;216:41–59.
[34] Hagemeyer N, Hanft K-M, Akriditou M-A, Unger N, Park ES, Stanley ER, et al. Microglia contribute to normal myelinogenesis and to oligodendrocyte progenitor maintenance during adulthood. Acta Neuropathol (Berl) 2017;134:441–58.
[35] Wlodarczyk A, Holtman IR, Krueger M, Yogev N, Bruttger J, Khorooshi R, et al. A novel microglial subset plays a key role in myelinogenesis in developing brain. EMBO J 2017;36:3292–308.
[36] Hubel DH. Exploration of the primary visual cortex, 1955–78. Nature 1982;299:515–24.
[37] Nelson CA, Fox NA, Zeanah CH. Romania's abandoned children: deprivation, brain development, and the struggle for recovery. Cambridge, MA: Harvard University Press; 2014.
[38] Pollak SD, Nelson CA, Schlaak MF, Roeber BJ, Wewerka SS, Wiik KL, et al. Neurodevelopmental effects of early deprivation in post-institutionalized children. Child Dev 2010;81:224–36.
[39] Nelson CA, Zeanah CH, Fox NA, Marshall PJ, Smyke AT, Guthrie D. Cognitive recovery in socially deprived young children: the Bucharest Early Intervention Project. Science 2007;318:1937–40.
[40] Bravo JA, Forsythe P, Chew MV, Escaravage E, Savignac HM, Dinan TG, et al. Ingestion of Lactobacillus strain regulates emotional behavior and central GABA receptor expression in a mouse via the vagus nerve. Proc Natl Acad Sci USA 2011;108:16050–5.
[41] Desbonnet L, Clarke G, Shanahan F, Dinan TG, Cryan JF. Microbiota is essential for social development in the mouse. Mol Psychiatry 2014;19:146–8.
[42] Gareau MG, Wine E, Rodrigues DM, Cho JH, Whary MT, Philpott DJ, et al. Bacterial infection causes stress-induced memory dysfunction in mice. Gut 2011;60:307–17.
[43] Lu J, Synowiec S, Lu L, Yu Y, Bretherick T, Takada S, et al. Microbiota influence the development of the brain and behaviors in C57BL/6J mice. PLoS One 2018;13:e0201829.

[44] Neufeld K-AM, Kang N, Bienenstock J, Foster JA. Effects of intestinal microbiota on anxiety-like behavior. Commun Integr Biol 2011;4:492–4.

[45] Braniste V, Al-Asmakh M, Kowal C, Anuar F, Abbaspour A, Tóth M, et al. The gut microbiota influences blood-brain barrier permeability in mice. Sci Transl Med 2014;6 263ra158-263ra158.

[46] Lu J, Lu L, Yu Y, Cluette-Brown J, Martin CR, Claud EC. Effects of intestinal microbiota on brain development in humanized gnotobiotic mice. Sci Rep 2018;8.

[47] Pan J-X, Deng F-L, Zeng B-H, Zheng P, Liang W-W, Yin B-M, et al. Absence of gut microbiota during early life affects anxiolytic behaviors and monoamine neurotransmitters system in the hippocampal of mice. J Neurol Sci 2019;400:160–8.

[48] Sudo N, Chida Y, Aiba Y, Sonoda J, Oyama N, Yu X-N, et al. Postnatal microbial colonization programs the hypothalamic–pituitary–adrenal system for stress response in mice. J Physiol 2004;558:263–75.

[49] Dam SA, Mostert JC, Szopinska-Tokov JW, Bloemendaal M, Amato M, Arias-Vasquez A. The role of the gut-brain axis in attention-deficit/hyperactivity disorder. Gastroenterol Clin North Am 2019;48:407–31.

[50] Pulikkan J, Mazumder A, Grace T. Role of the gut microbiome in autism spectrum disorders. Adv Exp Med Biol 2019;1118:253–69.

[51] Yuan X, Kang Y, Zhuo C, Huang X-F, Song X. The gut microbiota promotes the pathogenesis of schizophrenia via multiple pathways. Biochem Biophys Res Commun 2019;512:373–80.

[52] Zhu F, Guo R, Wang W, Ju Y, Wang Q, Ma Q, et al. Transplantation of microbiota from drug-free patients with schizophrenia causes schizophrenia-like abnormal behaviors and dysregulated kynurenine metabolism in mice. Mol Psychiatry 2019. Available from: https://doi.org/10.1038/s41380-019-0475-4.

[53] Cerdó T, Diéguez E, Campoy C. Early nutrition and gut microbiome: interrelationship between bacterial metabolism, immune system, brain structure, and neurodevelopment. Am J Physiol Endocrinol Metab 2019;317:E617–30.

[54] Farshim P, Walton G, Chakrabarti B, Givens I, Saddy D, Kitchen I, et al. Maternal weaning modulates emotional behavior and regulates the gut-brain axis. Sci Rep 2016;6:21958.

[55] Gur TL, Palkar AV, Rajasekera T, Allen J, Niraula A, Godbout J, et al. Prenatal stress disrupts social behavior, cortical neurobiology and commensal microbes in adult male offspring. Behav Brain Res 2019;359:886–94.

[56] Keunen K, van Elburg RM, van Bel F, Benders MJNL. Impact of nutrition on brain development and its neuroprotective implications following preterm birth. Pediatr Res 2015;77:148–55.

[57] O'Mahony SM, Marchesi JR, Scully P, Codling C, Ceolho A-M, Quigley EMM, et al. Early life stress alters behavior, immunity, and microbiota in rats: implications for irritable bowel syndrome and psychiatric illnesses. Biol Psychiatry 2009;65:263–7.

[58] Stewart CJ, Ajami NJ, O'Brien JL, Hutchinson DS, Smith DP, Wong MC, et al. Temporal development of the gut microbiome in early childhood from the TEDDY study. Nature 2018;562:583–8.

[59] Slykerman RF, Coomarasamy C, Wickens K, Thompson JMD, Stanley TV, Barthow C, et al. Exposure to antibiotics in the first 24 months of life and neurocognitive outcomes at 11 years of age. Psychopharmacology (Berl) 2019;236:1573–82.

[60] Slykerman RF, Thompson J, Waldie KE, Murphy R, Wall C, Mitchell EA. Antibiotics in the first year of life and subsequent neurocognitive outcomes. Acta Paediatr 2017;106:87–94.

[61] Lu L, Yu Y, Guo Y, Wang Y, Chang EB, Claud EC. Transcriptional modulation of intestinal innate defense/inflammation genes by preterm infant microbiota in a humanized gnotobiotic mouse model. PLoS One 2015;10:e0124504.
[62] Sharon G, Cruz NJ, Kang D-W, Gandal MJ, Wang B, Kim Y-M, et al. Human gut microbiota from autism spectrum disorder promote behavioral symptoms in mice. Cell 2019;177:1600–1618.e17.
[63] Raichle ME, Gusnard DA. Appraising the brain's energy budget. Proc Natl Acad Sci USA 2002;99:10237–9.
[64] Goyal MS, Venkatesh S, Milbrandt J, Gordon JI, Raichle ME. Feeding the brain and nurturing the mind: linking nutrition and the gut microbiota to brain development. Proc Natl Acad Sci USA 2015;112:14105–12.
[65] Goyal MS, Hawrylycz M, Miller JA, Snyder AZ, Raichle ME. Aerobic glycolysis in the human brain is associated with development and neotenous gene expression. Cell Metab 2014;19:49–57.
[66] Tseng KY, Chambers RA, Lipska BK. The neonatal ventral hippocampal lesion as a heuristic neurodevelopmental model of schizophrenia. Behav Brain Res 2009;204:295–305.
[67] Prado EL, Dewey KG. Nutrition and brain development in early life. Nutr Rev 2014;72:267–84.
[68] Morrison DJ, Preston T. Formation of short chain fatty acids by the gut microbiota and their impact on human metabolism. Gut Microbes 2016;7:189–200.
[69] den Besten G, van Eunen K, Groen AK, Venema K, Reijngoud D-J, Bakker BM. The role of short-chain fatty acids in the interplay between diet, gut microbiota, and host energy metabolism. J Lipid Res 2013;54:2325–40.
[70] Soto M, Herzog C, Pacheco JA, Fujisaka S, Bullock K, Clish CB, et al. Gut microbiota modulate neurobehavior through changes in brain insulin sensitivity and metabolism. Mol Psychiatry 2018;23:2287–301.
[71] Hallschmid M, Benedict C, Schultes B, Born J, Kern W. Obese men respond to cognitive but not to catabolic brain insulin signaling. Int J Obes 2008;32:275–82.
[72] Kleinridders A, Cai W, Cappellucci L, Ghazarian A, Collins WR, Vienberg SG, et al. Insulin resistance in brain alters dopamine turnover and causes behavioral disorders. Proc Natl Acad Sci USA 2015;112:3463–8.
[73] Navarro G, Sharma A, Dugas LR, Forrester T, Gilbert JA, Layden BT. Gut microbial features can predict host phenotype response to protein deficiency. Physiol Rep 2018;6:e13932.
[74] Portune KJ, Beaumont M, Davila A-M, Tomé D, Blachier F, Sanz Y. Gut microbiota role in dietary protein metabolism and health-related outcomes: the two sides of the coin. Trends Food Sci Technol 2016;57:213–32.
[75] Chang EB, Martinez-Guryn K. Small intestinal microbiota: the neglected stepchild needed for fat digestion and absorption. Gut Microbes 2018;10:235–40.
[76] Staley C, Weingarden AR, Khoruts A, Sadowsky MJ. Interaction of gut microbiota with bile acid metabolism and its influence on disease states. Appl Microbiol Biotechnol 2017;101:47–64.
[77] Biesalski HK. Nutrition meets the microbiome: micronutrients and the microbiota. Ann N Y Acad Sci 2016;1372:53–64.
[78] Romano KA, Vivas EI, Amador-Noguez D, Rey FE. Intestinal microbiota composition modulates choline bioavailability from diet and accumulation of the proatherogenic metabolite trimethylamine-N-oxide. mBio 2015;6:e02481.

[79] Skolnick SD, Greig NH. Microbes and monoamines: potential neuropsychiatric consequences of dysbiosis. Trends Neurosci 2019;42:151–63.
[80] Chang C-S, Kao C-Y. Current understanding of the gut microbiota shaping mechanisms. J Biomed Sci 2019;26:59.
[81] Subramanian S, Huq S, Yatsunenko T, Haque R, Mahfuz M, Alam MA, et al. Persistent gut microbiota immaturity in malnourished Bangladeshi children. Nature 2014;510:417–21.
[82] Younge NE, Newgard CB, Cotten CM, Goldberg RN, Muehlbauer MJ, Bain JR, et al. Disrupted maturation of the microbiota and metabolome among extremely preterm infants with postnatal growth failure. Sci Rep 2019;9:1–12.
[83] Gehrig JL, Venkatesh S, Chang H-W, Hibberd MC, Kung VL, Cheng J, et al. Effects of microbiota-directed foods in gnotobiotic animals and undernourished children. Science 2019;365:eaau4732.
[84] Wikoff WR, Anfora AT, Liu J, Schultz PG, Lesley SA, Peters EC, et al. Metabolomics analysis reveals large effects of gut microflora on mammalian blood metabolites. Proc Natl Acad Sci USA 2009;106:3698–703.
[85] Arentsen T, Qian Y, Gkotzis S, Femenia T, Wang T, Udekwu K, et al. The bacterial peptidoglycan-sensing molecule Pglyrp2 modulates brain development and behavior. Mol Psychiatry 2017;22:257–66.
[86] Bachmann C, Colombo JP, Berüter J. Short chain fatty acids in plasma and brain: quantitative determination by gas chromatography. Clin Chim Acta 1979;92:153–9.
[87] Caspani G, Swann J. Small talk: microbial metabolites involved in the signaling from microbiota to brain. Curr Opin Pharmacol 2019;48:99–106.
[88] Chu C, Murdock MH, Jing D, Won TH, Chung H, Kressel AM, et al. The microbiota regulate neuronal function and fear extinction learning. Nature 2019;574:543–8.
[89] Oldendorf WH. Carrier-mediated blood-brain barrier transport of short-chain monocarboxylic organic acids. Am J Physiol 1973;224:1450–3.
[90] Vargas-Caraveo A, Sayd A, Maus SR, Caso JR, Madrigal JLM, García-Bueno B, et al. Lipopolysaccharide enters the rat brain by a lipoprotein-mediated transport mechanism in physiological conditions. Sci Rep 2017;7:1–15.
[91] Citraro R, Leo A, De Caro C, Nesci V, Gallo Cantafio ME, Amodio N, et al. Effects of histone deacetylase inhibitors on the development of epilepsy and psychiatric comorbidity in WAG/Rij rats. Mol Neurobiol 2020;57:408–21.
[92] Reddy SD, Clossen BL, Reddy DS. Epigenetic histone deacetylation inhibition prevents the development and persistence of temporal lobe epilepsy. J Pharmacol Exp Ther 2018;364:97–109.
[93] Albuquerque Filho MO, de Freitas BS, Garcia RCL, Crivelaro PC de F, Schröder N, de Lima MNM. Dual influences of early-life maternal deprivation on histone deacetylase activity and recognition memory in rats. Neuroscience 2017;344:360–70.
[94] Val-Laillet D, Guérin S, Coquery N, Nogret I, Formal M, Romé V, et al. Oral sodium butyrate impacts brain metabolism and hippocampal neurogenesis, with limited effects on gut anatomy and function in pigs. FASEB J 2018;32:2160–71.
[95] Danilova AB, Grinkevich LN. Failure of long-term memory formation in juvenile snails is determined by acetylation status of histone H3 and can be improved by NaB treatment. PLoS One 2012;7:e41828.
[96] Shah P, Nankova BB, Parab S, La Gamma EF. Short chain fatty acids induce TH gene expression via ERK-dependent phosphorylation of CREB protein. Brain Res 2006;1107:13–23.

[97] Huang J, Zhang L, Qu Y, Zhou Y, Zhu J, Li Y, et al. Histone acetylation of oligodendrocytes protects against white matter injury induced by inflammation and hypoxia-ischemia through activation of BDNF-TrkB signaling pathway in neonatal rats. Brain Res 2018;1688:33–46.
[98] Kanungo MS, Thakur MK. Modulation of acetylation of histones and transcription of chromatin by butyric acid and 17β-estradiol in the brain of rats of various ages. Biochem Biophys Res Commun 1979;87:266–71.
[99] Nankova BB, Agarwal R, MacFabe DF, Gamma EFL. Enteric bacterial metabolites propionic and butyric acid modulate gene expression, including CREB-dependent catecholaminergic neurotransmission, in PC12 cells – possible relevance to autism spectrum disorders. PLoS One 2014;9:e103740.
[100] Shultz SR, Aziz NAB, Yang L, Sun M, MacFabe DF, O'Brien TJ. Intracerebroventricular injection of propionic acid, an enteric metabolite implicated in autism, induces social abnormalities that do not differ between seizure-prone (FAST) and seizure-resistant (SLOW) rats. Behav Brain Res 2015;278:542–8.
[101] Hoyles L, Snelling T, Umlai U-K, Nicholson JK, Carding SR, Glen RC, et al. Microbiome–host systems interactions: protective effects of propionate upon the blood–brain barrier. Microbiome 2018;6:55.
[102] Frost G, Sleeth ML, Sahuri-Arisoylu M, Lizarbe B, Cerdan S, Brody L, et al. The short-chain fatty acid acetate reduces appetite via a central homeostatic mechanism. Nat Commun 2014;5:1–11.
[103] Paolicelli RC, Bolasco G, Pagani F, Maggi L, Scianni M, Panzanelli P, et al. Synaptic pruning by microglia is necessary for normal brain development. Science 2011;333:1456–8.
[104] Keşli R, Gökçen C, Buluğ U, Terzi Y. Investigation of the relation between anaerobic bacteria genus *Clostridium* and late-onset autism etiology in children. J Immunoassay Immunochem 2014;35:101–9.
[105] Shaw W. Increased urinary excretion of a 3-(3-hydroxyphenyl)-3-hydroxypropionic acid (HPHPA), an abnormal phenylalanine metabolite of *Clostridia* spp. in the gastrointestinal tract, in urine samples from patients with autism and schizophrenia. Nutr Neurosci 2010;13:135–43.
[106] Clayton TA. Metabolic differences underlying two distinct rat urinary phenotypes, a suggested role for gut microbial metabolism of phenylalanine and a possible connection to autism. FEBS Lett 2012;586:956–61.
[107] Hsiao EY, McBride SW, Hsien S, Sharon G, Hyde ER, McCue T, et al. Microbiota modulate behavioral and physiological abnormalities associated with neurodevelopmental disorders. Cell 2013;155:1451–63.
[108] Carregosa D, Carecho R, Figueira I, N Santos C. Low-molecular weight metabolites from polyphenols as effectors for attenuating neuroinflammation. J Agric Food Chem 2020;68:1790–807.
[109] Westfall S, Pasinetti GM. The gut microbiota links dietary polyphenols with management of psychiatric mood disorders. Front Neurosci 2019;13.
[110] Gao J, Xu K, Liu H, Liu G, Bai M, Peng C, et al. Impact of the gut microbiota on intestinal immunity mediated by tryptophan metabolism. Front Cell Infect Microbiol 2018;8.
[111] Latchney SE, Hein AM, O'Banion MK, DiCicco-Bloom E, Opanashuk LA. Deletion or activation of the aryl hydrocarbon receptor alters adult hippocampal neurogenesis and contextual fear memory. J Neurochem 2013;125:430–45.

[112] Islam J, Sato S, Watanabe K, Watanabe T, Ardiansyah, Hirahara K, et al. Dietary tryptophan alleviates dextran sodium sulfate-induced colitis through aryl hydrocarbon receptor in mice. J Nutr Biochem 2017;42:43–50.
[113] Tourino MC, de Oliveira EM, Bellé LP, Knebel FH, Albuquerque RC, Dörr FA, et al. Tryptamine and dimethyltryptamine inhibit indoleamine 2,3 dioxygenase and increase the tumor-reactive effect of peripheral blood mononuclear cells. Cell Biochem Funct 2013;31:361–4.
[114] Rothhammer V, Borucki DM, Tjon EC, Takenaka MC, Chao C-C, Fabregat AA, et al. Microglial control of astrocytes in response to microbial metabolites. Nature 2018;557:724–8.
[115] Kim JW, Hong JY, Bae SM. Microglia and autism spectrum disorder: overview of current evidence and novel immunomodulatory treatment options. Clin Psychopharmacol Neurosci 2018;16:246–52.
[116] Wang M, Zhang L, Gage FH. Microglia, complement and schizophrenia. Nat Neurosci 2019;22:333–4.
[117] Greenwood J, Adu J, Davey AJ, Abbott NJ, Bradbury MW. The effect of bile salts on the permeability and ultrastructure of the perfused, energy-depleted, rat blood-brain barrier. J Cereb Blood Flow Metab 1991;11:644–54.
[118] Palmela I, Correia L, Silva RFM, Sasaki H, Kim KS, Brites D, et al. Hydrophilic bile acids protect human blood-brain barrier endothelial cells from disruption by unconjugated bilirubin: an in vitro study. Front Neurosci 2015;9.
[119] Quinn M, McMillin M, Galindo C, Frampton G, Pae HY, DeMorrow S. Bile acids permeabilize the blood brain barrier after bile duct ligation in rats via Rac1-dependent mechanisms. Dig Liver Dis 2014;46:527–34.
[120] Reddy IA, Smith NK, Erreger K, Ghose D, Saunders C, Foster DJ, et al. Bile diversion, a bariatric surgery, and bile acid signaling reduce central cocaine reward. PLoS Biol 2018;16:e2006682.
[121] Money KM, Stanwood GD. Developmental origins of brain disorders: roles for dopamine. Front Cell Neurosci 2013;7.
[122] Heiss CN, Olofsson LE. The role of the gut microbiota in development, function and disorders of the central nervous system and the enteric nervous system. J Neuroendocrinol 2019;31.
[123] Forsythe P, Bienenstock J, Kunze WA. Vagal pathways for microbiome-brain-gut axis communication. In: Lyte M, Cryan JF, editors. Microbial endocrinology: the microbiota-gut-brain axis in health and disease, advances in experimental medicine and biology. New York: Springer; 2014. p. 115–33.
[124] Bonaz B, Bazin T, Pellissier S. The vagus nerve at the interface of the microbiota-gut-brain axis. Front Neurosci 2018;12.
[125] Walsh KT, Zemper AE. The enteric nervous system for epithelial researchers: basic anatomy, techniques, and interactions with the epithelium. Cell Mol Gastroenterol Hepatol 2019;8:369–78.
[126] Contestabile A. Roles of NMDA receptor activity and nitric oxide production in brain development. Brain Res Rev 2000;32:476–509.
[127] du Bois TM, Huang X-F. Early brain development disruption from NMDA receptor hypofunction: relevance to schizophrenia. Brain Res Rev 2007;53:260–70.
[128] Rezvani AH. Involvement of the NMDA system in learning and memory. In: Levin ED, Buccafusco JJ, editors. Animal models of cognitive impairment, frontiers in neuroscience. Boca Raton, FL: CRC Press/Taylor & Francis; 2006.

[129] Ates-Alagoz Z, Adejare A. NMDA receptor antagonists for treatment of depression. Pharmaceuticals 2013;6:480–99.

[130] Bubeníková-Valesová V, Balcar VJ, Tejkalová H, Langmeier M, St'astný F. Neonatal administration of N-acetyl-L-aspartyl-L-glutamate induces early neurodegeneration in hippocampus and alters behaviour in young adult rats. Neurochem Int 2006;48:515–22.

[131] Neufeld KM, Kang N, Bienenstock J, Foster JA. Reduced anxiety-like behavior and central neurochemical change in germ-free mice. Neurogastroenterol Motil 2011;23:255–e119.

[132] Strandwitz P, Kim KH, Terekhova D, Liu JK, Sharma A, Levering J, et al. GABA modulating bacteria of the human gut microbiota. Nat Microbiol 2019;4:396–403.

[133] Maqsood R, Stone TW. The gut-brain axis, BDNF, NMDA and CNS disorders. Neurochem Res 2016;41:2819–35.

[134] Fernandes BS, Steiner J, Berk M, Molendijk ML, Gonzalez-Pinto A, Turck CW, et al. Peripheral brain-derived neurotrophic factor in schizophrenia and the role of antipsychotics: meta-analysis and implications. Mol Psychiatry 2015;20:1108–19.

[135] Lindholm JSO, Castrén E. Mice with altered BDNF signaling as models for mood disorders and antidepressant effects. Front Behav Neurosci 2014;8.

[136] Saghazadeh A, Rezaei N. Brain-derived neurotrophic factor levels in autism: a systematic review and meta-analysis. J Autism Dev Disord 2017;47:1018–29.

[137] Weidner KL, Buenaventura DF, Chadman KK. Mice over-expressing BDNF in forebrain neurons develop an altered behavioral phenotype with age. Behav Brain Res 2014;268:222–8.

[138] O'Sullivan E, Barrett E, Grenham S, Fitzgerald P, Stanton C, Ross RP, et al. BDNF expression in the hippocampus of maternally separated rats: does *Bifidobacterium* breve 6330 alter BDNF levels? Benef Microbes 2011;2:199–207.

[139] Wei Y, Melas PA, Wegener G, Mathé AA, Lavebratt C. Antidepressant-like effect of sodium butyrate is associated with an increase in TET1 and in 5-hydroxymethylation levels in the Bdnf gene. Int J Neuropsychopharmacol 2014;18.

[140] Yano JM, Yu K, Donaldson GP, Shastri GG, Ann P, Ma L, et al. Indigenous bacteria from the gut microbiota regulate host serotonin biosynthesis. Cell 2015;161:264–76.

[141] Shah R, Courtiol E, Castellanos FX, Teixeira CM. Abnormal serotonin levels during perinatal development lead to behavioral deficits in adulthood. Front Behav Neurosci 2018;12.

[142] Chameau P, Inta D, Vitalis T, Monyer H, Wadman WJ, van Hooft JA. The N-terminal region of reelin regulates postnatal dendritic maturation of cortical pyramidal neurons. Proc Natl Acad Sci USA 2009;106:7227–32.

[143] Kepser L-J, Homberg JR. The neurodevelopmental effects of serotonin: a behavioural perspective. Behav Brain Res 2015;277:3–13 Spec Issue: Serotonin.

[144] Lee L-J. Neonatal fluoxetine exposure affects the neuronal structure in the somatosensory cortex and somatosensory-related behaviors in adolescent rats. Neurotox Res 2009;15:212–23.

[145] Lee L-J, Lee LJ-H. Neonatal fluoxetine exposure alters motor performances of adolescent rats. Dev Neurobiol 2012;72:1122–32.

[146] Miceli S, Negwer M, van Eijs F, Kalkhoven C, van Lierop I, Homberg J, et al. High serotonin levels during brain development alter the structural input-output connectivity of neural networks in the rat somatosensory layer IV. Front Cell Neurosci 2013;7.

[147] Bellono NW, Bayrer JR, Leitch DB, Castro J, Zhang C, O'Donnell T, et al. Enterochromaffin cells are gut chemosensors that couple to sensory neural pathways. Cell 2017;170:185–198.e16.

[148] Sjögren K, Engdahl C, Henning P, Lerner UH, Tremaroli V, Lagerquist MK, et al. The gut microbiota regulates bone mass in mice. J Bone Miner Res 2012;27:1357–67.

[149] Andersen SL, Teicher MH. Delayed effects of early stress on hippocampal development. Neuropsychopharmacology 2004;29:1988–93.

[150] Wei L, Hao J, Lacher RK, Abbott T, Chung L, Colangelo CM, et al. Early-life stress perturbs key cellular programs in the developing mouse hippocampus. Dev Neurosci 2015;37:476–88.

[151] Diaz Heijtz R, Wang S, Anuar F, Qian Y, Björkholm B, Samuelsson A, et al. Normal gut microbiota modulates brain development and behavior. Proc Natl Acad Sci USA 2011;108:3047–52.

[152] Turnbull AV, Rivier CL. Regulation of the hypothalamic-pituitary-adrenal axis by cytokines: actions and mechanisms of action. Physiol Rev 1999;79:1–71.

[153] Barcik W, Wawrzyniak M, Akdis CA, O'Mahony L. Immune regulation by histamine and histamine-secreting bacteria. Curr Opin Immunol 2017;48:108–13.

[154] Baronio D, Gonchoroski T, Castro K, Zanatta G, Gottfried C, Riesgo R. Histaminergic system in brain disorders: lessons from the translational approach and future perspectives. Ann Gen Psychiatry 2014;13:34.

[155] Bernardino L, Eiriz MF, Santos T, Xapelli S, Grade S, Rosa AI, et al. Histamine stimulates neurogenesis in the rodent subventricular zone. Stem Cells 2012;30:773–84.

[156] Rangon C-M, Schang A-L, Van Steenwinckel J, Schwendimann L, Lebon S, Fu T, et al. Myelination induction by a histamine H3 receptor antagonist in a mouse model of preterm white matter injury. Brain Behav Immun 2018;74:265–76.

[157] Elenkov IJ, Webster E, Papanicolaou DA, Fleisher TA, Chrousos GP, Wilder RL. Histamine potently suppresses human IL-12 and stimulates IL-10 production via H2 receptors. J Immunol 1998;1950(161):2586–93.

[158] Thomas CM, Hong T, van Pijkeren JP, Hemarajata P, Trinh DV, Hu W, et al. Histamine derived from probiotic *Lactobacillus reuteri* suppresses TNF via modulation of PKA and ERK signaling. PLoS One 2012;7:e31951.

[159] Hart K, Kim L, Zaman M, Grimont C, D'Onofrio D, Freedman S, et al. Gut hormone levels in preterm infants as predictors of feeding intolerance. FASEB J 2015;29 901.23.

[160] Fawkner-Corbett D, Simmons A, Parikh K. Microbiome, pattern recognition receptor function in health and inflammation. Best Pract Res Clin Gastroenterol 2017;31:683–91.

[161] Postler TS, Ghosh S. Understanding the holobiont: how microbial metabolites affect human health and shape the immune system. Cell Metab 2017;26:110–30.

[162] Schirmer M, Smeekens SP, Vlamakis H, Jaeger M, Oosting M, Franzosa EA, et al. Linking the human gut microbiome to inflammatory cytokine production capacity. Cell 2016;167:1125–1136.e8.

[163] Stolp HB. Neuropoietic cytokines in normal brain development and neurodevelopmental disorders. Mol Cell Neurosci 2013;53:63–8.

[164] Greiner TU, Hyötyläinen T, Knip M, Bäckhed F, Orešič M. The gut microbiota modulates glycaemic control and serum metabolite profiles in non-obese diabetic mice. PLoS One 2014;9.

[165] Molinaro A, Caesar R, Holm LM, Tremaroli V, Cani PD, Bäckhed F. Host–microbiota interaction induces bi-phasic inflammation and glucose intolerance in mice. Mol Metab 2017;6:1371–80.

[166] Maddux AB, Douglas IS. Is the developmentally immature immune response in paediatric sepsis a recapitulation of immune tolerance? Immunology 2015;145:1–10.
[167] Hartman ME, Linde-Zwirble WT, Angus DC, Watson RS. Trends in the epidemiology of pediatric severe sepsis. Pediatr Crit Care Med 2013;14:686–93.
[168] Cardoso FL, Herz J, Fernandes A, Rocha J, Sepodes B, Brito MA, et al. Systemic inflammation in early neonatal mice induces transient and lasting neurodegenerative effects. J Neuroinflammation 2015;12.
[169] Dammann O, O'Shea TM. Cytokines and perinatal brain damage. Clin Perinatol 2008;35:643–63.
[170] Fowlie G, Cohen N, Ming X. The perturbance of microbiome and gut-brain axis in autism spectrum disorders. Int J Mol Sci 2018;19.
[171] Jiang NM, Tofail F, Moonah SN, Scharf RJ, Taniuchi M, Ma JZ, et al. Febrile illness and pro-inflammatory cytokines are associated with lower neurodevelopmental scores in Bangladeshi infants living in poverty. BMC Pediatr 2014;14:50.
[172] Rees CM, Pierro A, Eaton S. Neurodevelopmental outcomes of neonates with medically and surgically treated necrotizing enterocolitis. Arch Dis Child Fetal Neonatal Ed 2007;92:F193–8.
[173] Savioli (Fernandez) K, Rouse C, Gorman G, Susi A, Hisle-Gorman E. Suspected or known neonatal sepsis and neurodevelopmental outcomes at five years. Pediatrics 2018;141 760–760.
[174] Schulzke SM, Deshpande GC, Patole SK. Neurodevelopmental outcomes of very low-birth-weight infants with necrotizing enterocolitis: a systematic review of observational studies. Arch Pediatr Adolesc Med 2007;161:583–90.
[175] Jiang NM, Cowan M, Moonah SN, Petri WA. The impact of systemic inflammation on neurodevelopment. Trends Mol Med 2018;24:794–804.
[176] Lee SE, West KP, Cole RN, Schulze KJ, Wu LS-F, Yager JD, et al. General intelligence is associated with subclinical inflammation in Nepalese children: a population-based plasma proteomics study. Brain Behav Immun 2016;56:253–63.
[177] Osterholm EA, Georgieff MK. Chronic inflammation and iron metabolism. J Pediatr 2015;166:1351–1357.e1.
[178] Comito D, Romano C. Dysbiosis in the pathogenesis of pediatric inflammatory bowel diseases. Int J Inflamm 2012; https://www.hindawi.com/journals/iji/2012/687143/ (accessed 10.18.19).
[179] Neu J, Pammi M. Necrotizing enterocolitis: the intestinal microbiome, metabolome and inflammatory mediators. Semin Fetal Neonatal Med 2018;23:400–5.
[180] DiSabato D, Quan N, Godbout JP. Neuroinflammation: the devil is in the details. J Neurochem 2016;139:136–53.
[181] Moretti R, Pansiot J, Bettati D, Strazielle N, Ghersi-Egea J-F, Damante G, et al. Blood-brain barrier dysfunction in disorders of the developing brain. Front Neurosci 2015;9.
[182] Stolp HB, Johansson PA, Habgood MD, Dziegielewska KM, Saunders NR, Ek CJ. Effects of neonatal systemic inflammation on blood-brain barrier permeability and behaviour in juvenile and adult rats. Cardiovasc Psychiatry Neurol 2011;2011.
[183] Kealy J, Greene C, Campbell M. Blood-brain barrier regulation in psychiatric disorders. Neurosci Lett 2018;133664.
[184] Patra A, Huang H, Bauer JA, Giannone PJ. Neurological consequences of systemic inflammation in the premature neonate. Neural Regen Res 2017;12:890–6.

[185] Vela JM, Molina-Holgado E, Arévalo-Martín Á, Almazán G, Guaza C. Interleukin-1 regulates proliferation and differentiation of oligodendrocyte progenitor cells. Mol Cell Neurosci 2002;20:489–502.

[186] Lee SC, Liu W, Dickson DW, Brosnan CF, Berman JW. Cytokine production by human fetal microglia and astrocytes. Differential induction by lipopolysaccharide and IL-1 beta. J Immunol 1993;150:2659–67.

[187] Patel RK, Prasad N, Kuwar R, Haldar D, Abdul-Muneer PM. Transforming growth factor-beta 1 signaling regulates neuroinflammation and apoptosis in mild traumatic brain injury. Brain Behav Immun 2017;64:244–58.

[188] Falk S, Wurdak H, Ittner LM, Ille F, Sumara G, Schmid M-T, et al. Brain area-specific effect of TGF-β signaling on Wnt-dependent neural stem cell expansion. Cell Stem Cell 2008;2:472–83.

[189] Adachi T, Takanaga H, Kunimoto M, Asou H. Influence of LIF and BMP-2 on differentiation and development of glial cells in primary cultures of embryonic rat cerebral hemisphere. J Neurosci Res 2005;79:608–15.

[190] Stipursky J, Francis D, Dezonne RS, Bérgamo A de A, Souza L, Moraes CA, et al. TGF-β1 promotes cerebral cortex radial glia-astrocyte differentiation in vivo. Front Cell Neurosci 2014;8 393–393.

[191] Vargas DL, Nascimbene C, Krishnan C, Zimmerman AW, Pardo CA. Neuroglial activation and neuroinflammation in the brain of patients with autism. Ann Neurol 2005;57:67–81.

[192] Depino AM, Lucchina L, Pitossi F. Early and adult hippocampal TGF-β1 overexpression have opposite effects on behavior. Brain Behav Immun 2011;25:1582–91.

[193] Bauché D, Marie JC. Transforming growth factor β: a master regulator of the gut microbiota and immune cell interactions. Clin Transl Immunol 2017;6:e136.

[194] Sanford LP, Ormsby I, Gittenberger-de Groot AC, Sariola H, Friedman R, Boivin GP, et al. TGFbeta2 knockout mice have multiple developmental defects that are non-overlapping with other TGFbeta knockout phenotypes. Development 1997;124:2659–70.

[195] Back SA, Luo NL, Mallinson RA, O'Malley JP, Wallen LD, Frei B, et al. Selective vulnerability of preterm white matter to oxidative damage defined by F2-isoprostanes. Ann Neurol 2005;58:108–20.

[196] Haynes RL, Folkerth RD, Keefe RJ, Sung I, Swzeda LI, Rosenberg PA, et al. Nitrosative and oxidative injury to premyelinating oligodendrocytes in periventricular leukomalacia. J Neuropathol Exp Neurol 2003;62:441–50.

[197] Iismaa SE, Kaidonis X, Nicks AM, Bogush N, Kikuchi K, Naqvi N, et al. Comparative regenerative mechanisms across different mammalian tissues. NPJ Regen Med 2018;3:1–20.

[198] Hanisch U-K. Microglia as a source and target of cytokines. Glia 2002;40:140–55.

[199] Nakashima K, Wiese S, Yanagisawa M, Arakawa H, Kimura N, Hisatsune T, et al. Developmental requirement of gp130 signaling in neuronal survival and astrocyte differentiation. J Neurosci 1999;19:5429–34.

[200] Smith SEP, Li J, Garbett K, Mirnics K, Patterson PH. Maternal immune activation alters fetal brain development through interleukin-6. J Neurosci 2007;27:10695–702.

[201] Wei H, Chadman KK, McCloskey DP, Sheikh AM, Malik M, Brown WT, et al. Brain IL-6 elevation causes neuronal circuitry imbalances and mediates autism-like behaviors. Biochim Biophys Acta 2012;1822:831–42.

[202] Yanguas-Casás N, Barreda-Manso MA, Nieto-Sampedro M, Romero-Ramírez L. TUDCA: an agonist of the bile acid receptor GPBAR1/TGR5 with anti-inflammatory effects in microglial cells. J Cell Physiol 2017;232:2231–45.
[203] Ruiz PA, Hoffmann M, Szcesny S, Blaut M, Haller D. Innate mechanisms for *Bifidobacterium* lactis to activate transient pro-inflammatory host responses in intestinal epithelial cells after the colonization of germ-free rats. Immunology 2005;115:441–50.
[204] Chourbaji S, Urani A, Inta I, Sanchis-Segura C, Brandwein C, Zink M, et al. IL-6 knockout mice exhibit resistance to stress-induced development of depression-like behaviors. Neurobiol Dis 2006;23:587–94.
[205] Wallenius V, Wallenius K, Ahrén B, Rudling M, Carlsten H, Dickson SL, et al. Interleukin-6-deficient mice develop mature-onset obesity. Nat Med 2002;8:75–9.
[206] Zablotsky B, Black LI, Maenner MJ, Schieve LA, Danielson ML, Bitsko RH, et al. Prevalence and trends of developmental disabilities among children in the United States: 2009–2017. Pediatrics 2019;144:e20190811.
[207] Croen LA, Zerbo O, Qian Y, Massolo ML, Rich S, Sidney S, et al. The health status of adults on the autism spectrum. Autism 2015;19:814–23.
[208] Miot S, Akbaraly T, Michelon C, Couderc S, Crepiat S, Loubersac J, et al. Comorbidity burden in adults with autism spectrum disorders and intellectual disabilities—a report from the EFAAR (Frailty Assessment in Ageing Adults with Autism Spectrum and Intellectual Disabilities) Study. Front Psychiatry 2019;10.
[209] Cani PD. Human gut microbiome: hopes, threats and promises. Gut 2018;67:1716–25.
[210] Durack J, Lynch SV. The gut microbiome: relationships with disease and opportunities for therapy. J Exp Med 2019;216:20–40.
[211] Kho ZY, Lal SK. The human gut microbiome – a potential controller of wellness and disease. Front Microbiol 2018;9.
[212] Azad MB, Bridgman SL, Becker AB, Kozyrskyj AL. Infant antibiotic exposure and the development of childhood overweight and central adiposity. Int J Obes 2014;38:1290–8.
[213] Boursi B, Mamtani R, Haynes K, Yang Y-X. The effect of past antibiotic exposure on diabetes risk. Eur J Endocrinol 2015;172:639–48.
[214] Kronman MP, Zaoutis TE, Haynes K, Feng R, Coffin SE. Antibiotic exposure and IBD development among children: a population-based cohort study. Pediatrics 2012;130: e794–803.
[215] Risnes KR, Belanger K, Murk W, Bracken MB. Antibiotic exposure by 6 months and asthma and allergy at 6 years: findings in a cohort of 1,401 US children. Am J Epidemiol 2011;173:310–18.
[216] Shaw SY, Blanchard JF, Bernstein CN. Association between the use of antibiotics in the first year of life and pediatric inflammatory bowel disease. Am J Gastroenterol 2010;105:2687–92.

Chapter 7

Microbiome and food allergy

Stephanie Jeong[1], Hsi-en Ho[1] and Supinda Bunyavanich[1,2]
[1]Division of Allergy and Immunology, Department of Pediatrics, Icahn School of Medicine at Mount Sinai, New York, NY, United States, [2]Icahn Institute for Data Science and Genomic Technology, Department of Genetics & Genomic Sciences, Icahn School of Medicine at Mount Sinai, New York, NY, United States

Overview of food allergy

The prevalence of food allergy has increased dramatically in the last 2–3 decades [1]. It has become a major health problem, with an estimated 5% of adults and 8%–10% of children affected in some western/industrialized countries [1,2]. Food allergy is commonly defined as an adverse health effect arising from a specific immune response that occurs reproducibly on exposure to a given food [3]. The most common form of food allergies is one mediated by allergen-specific immunoglobulin (Ig) E [3]. A limited list of inciting foods accounts for the majority of disease burden, namely milk, egg, peanut, tree nuts, wheat, soy, fish, shellfish, and seeds [1]. Clinically, an IgE-mediated food-allergic reaction is typically rapid in onset, involving the skin (e.g., hives), respiratory (e.g., wheezing), gastrointestinal (e.g., vomiting), and/or cardiovascular systems (e.g., hypotension) [3]. Severity can range from mild local symptoms to life-threatening anaphylaxis. Food allergy significantly impacts the medical, nutritional, and psychosocial health of affected individuals and has become an area of significant research efforts [1].

The development of food allergies is thought to be a result of a breakdown of tolerance to food products (Fig. 7.1) [4]. In a healthy individual, oral tolerance is the default response to ingested food antigens, and this is an active immune process. It begins with the digestion and uptake of food antigens in the small intestines, followed by the initiation of immune response at the gut-associated lymphoid tissues (GALT) in the regional lymph nodes [4]. This multistep process involves the capturing and processing of antigens by resident antigen-presenting cells, antigen presentation and initiation of T-cell response in the lymphoid tissues, and formation of effector immune cells [4]. In a healthy individual, this T-cell response is thought to be characterized by

The Developing Microbiome. DOI: https://doi.org/10.1016/B978-0-12-820602-7.00007-6

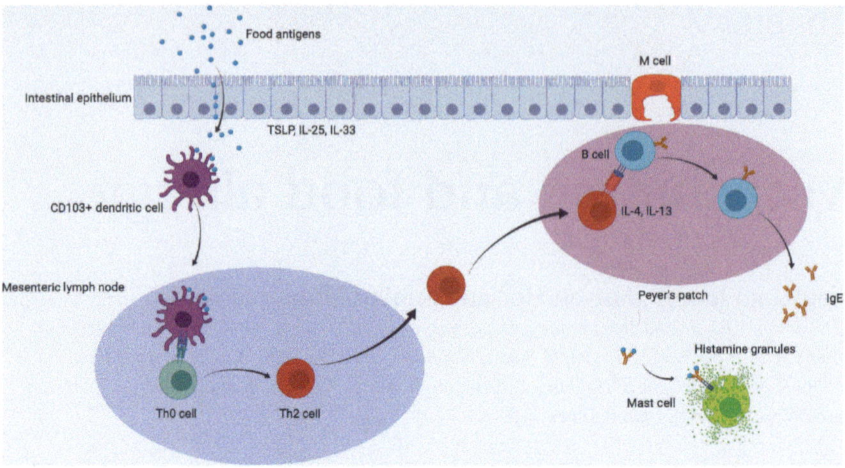

FIGURE 7.1 Development of food allergy. Food is digested in the gastrointestinal tract and food antigens pass through the intestinal epithelium. Food antigens can be taken up by antigen-presenting cells (APCs) such as dendritic cells, which can reach between intestinal epithelial cells and take up antigens directly, as well as M cells, which reside in Peyer's patches in the intestine and deliver antigens to APCs. Epithelial cells also produce cytokines such as thymic stromal lymphopoietin (TSLP), IL-25, and IL-33, which stimulate dendritic cells. CD103 + dendritic cells then take up antigen in the lamina propria and deliver them to mesenteric lymph nodes, where they present these peptides to naïve CD4 + T-cells (Th0 cells). This provides the stimulation required to promote differentiation into T helper type 2 (Th2) cells, which in turn produces IL-4 and IL-13, and provides stimulation through the CD40L/CD40 complex to induce B cells to undergo class switch and produces IgE antibodies that are specific to the antigen. When these IgE antibodies encounter the antigen again, they bind to mast cells via the FcεRI receptor, causing degranulation and release of histamine, triggering allergic symptoms. Figure created with BioRender.com.

the development of regulatory T (Treg) cells, which actively suppresses: (1) food-specific IgE production, (2) generation of pathogenic T helper type 2 (Th2) cells, and (3) mast cell activation, resulting in a state of oral tolerance [4]. In food-allergic individuals, there is a breakdown of immunological and clinical tolerance to an ingested food antigen. While the precise mechanism remains to be fully elucidated, the generation of Treg cells is believed to be compromised and a skewed Th2 response becomes prominent [4]. Th2 cells and associated cytokines induce antibody class-switch recombination toward IgEs, which then facilitate adverse reactions in allergic individuals. Upon exposure to food allergens, food-specific IgE antibodies bind to their receptors on the surface of mast cells and basophils, resulting in cell degranulation of preformed molecules, such as histamine, as well as rapid synthesis of lipid mediators, cytokines and chemokines, leading to aforementioned clinical symptoms [4]. The development of food-specific IgE is therefore considered to be the hallmark of allergen sensitization.

Epidemiologic associations between microbial exposure risk factors and allergy

A potential role for microbial exposure in allergy risk was prompted in the late 1980s with the observation that children from larger households tended to have lower rates of allergic rhinitis and eczema [5]. This contributed to the hygiene hypothesis, which posits that exposure to a variety of microorganisms is critical to the development of an appropriate immune response [5].

Subsequent findings from epidemiologic studies have suggested that early microbial exposure can mediate the risk of atopy, including food allergy in a number of cases [6]. For instance, having older siblings and dog ownership have been linked to lower risk for egg allergy development by 1 year of age [7]. In a number of birth cohort studies, cesarean delivery was associated with higher risk of food allergy development as well [8]. These environmental exposure factors have all been linked to differential microbial exposure during early childhood [6]. Comparison of genetically similar populations between Eastern and Western Europe has also provided support to the importance of the interaction between the environment, host microbiome, and food allergy. For instance, the gut microbiota of infants from Eastern Europe, where the prevalence of atopy was low, and Western Europe, where it was high, have been reported to be distinct [9,10]. Interestingly, the microbiota of the Eastern European infants shared similarities with those of Swedish infants three decades prior, suggesting a shift in the microbiota of infants from industrialized countries over recent decades [9,10]. Furthermore, the gut microbiota composition between children with food sensitization from both Western and Eastern Europe has also been found to be distinct from those without atopic diseases from these geographical regions [9,10].

The gastrointestinal microbiome and food allergy: findings from murine models

The microbiome as a mediator of food allergy is supported by murine models of food allergy [6,11]. For instance, in a food allergy-prone murine model (gain-of-function mutation in the IL-4 receptor α chain), there is a significant decrease in the relative abundance of bacteria in the Firmicutes family, and there is a significant increase in the relative abundance of bacteria in the Proteobacteria family, compared to wild-type mice [12]. Interestingly, the food allergy phenotype could be transmitted by transferring the gut flora from the allergy-prone mice to wild-type mice [12]. This important finding suggests that disease-associated microbiota may have a pathogenic role in the development of food allergy. Conversely, colonization with fecal material from a healthy donor can also be protective against development of food allergy in murine models [13,14].

Commensal bacteria have been found to influence the regulatory tone of the host immune system in murine models [15]. For instance, microbial colonization has been shown to promote the expansion of Treg cells. Specifically, the presence of microbiota induces a subset of RoRγt+ Treg cells [16]. In the absence of such RoRγt+ Treg cells, pathologic T helper type 2 response is exacerbated [16]. Numerous pathways in which microbial signals may influence oral tolerance have been elucidated in murine models (Fig. 7.2) [15]. Microbial signals can promote local production of retinoid and IL-10, which in turn can promote the expansion of local Treg cells [17]. Other bacterial signals have also been shown to stimulate TGF-β secretion from intestinal epithelial cells, which subsequently promote colonic Treg generation [18]. Lastly, signals from specific bacteria can induce the expression of key antiinflammatory molecules, such as IL-10 and inducible T-cell costimulator (ICOS), in Treg cells, resulting in an attenuated allergic phenotype [18].

One of the key observations in animal models of food allergy is that the protective effects of a healthy microbiome may be conferred by specific bacteria. For instance, oral administration of Clostridia strains to mice led to the

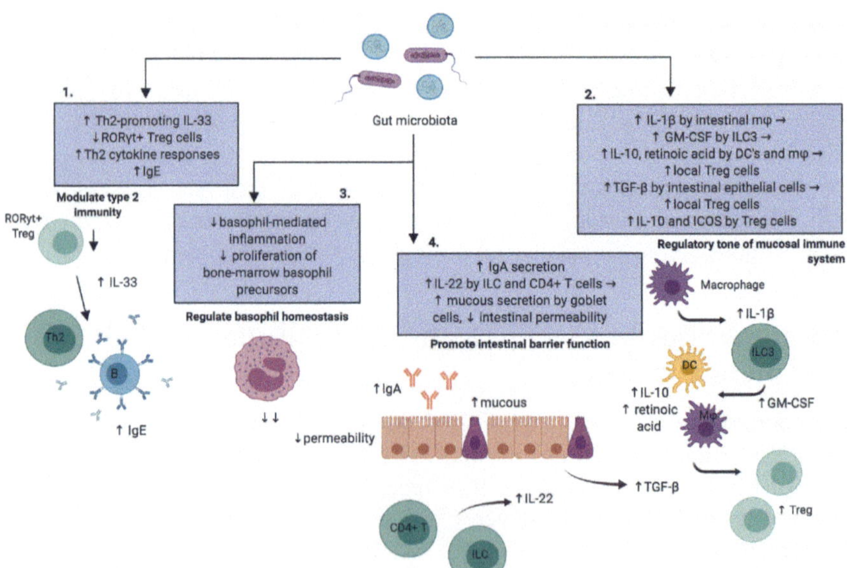

FIGURE 7.2 Mechanisms by which gut microbiota modulate food allergy susceptibility or oral tolerance. Murine models of food allergy have shown that gut microbiota are involved in regulating the intestinal mucosa and immunity. Gut microbiota (1) modulate type 2 immunity, (2) maintain the regulatory tone of the mucosal immune system, (3) regulate basophil homeostasis, and (4) promote intestinal barrier function. *Adapted from Ho HE, Bunyavanich S. Role of the microbiome in food allergy. Curr Allergy Asthma Rep 2018;18:27. Figure created with BioRender.com.*

induction of regulatory T-cells to attenuate models of colitis and allergic diarrhea [18,19]. Colonization with *Clostridium*, but not *Lactobacillus* or segmented filamentous bacteria, led to the induction of colonic Treg cells [19]. A follow-up study of mice colonized with feces from a healthy human donor demonstrated 17 strains of bacteria, all Clostridia detected in these mice by 16S rRNA gene sequencing [19]. These 17 strains were then administered to adult-specific pathogen-free (SPF) mice and were subjected to an ovalbumin (OVA)-induced allergic diarrhea model, and the 17-mix mice had significantly decreased occurrence and severity of allergic diarrhea as compared to SPF control mice ($P < .01$ at 3 weeks) [18]. Furthermore, when these mice were subjected to an experimental colitis model, the SPF + 17-mix mice showed less severe histological disease features and colon shortening, as well as lower mortality than control mice [18].

Other strains of bacteria, including *Bacillus*, *Lactobaccillus*, *Lactococcus*, and *Pediococcus*, have been implicated in the prevention of food allergies in mice as well. Germ-free mice colonized with *Bacteroides fragilis* had higher levels of antiinflammatory cytokine production than germ-free control mice [20]. A probiotic mixture of *Lactobacillus pentosus* KF340, *Lactobacillus paracasei* 698, *Lactococcus lactis* KF140, *Pediococcus pentosaceus* KF159, and *Bacillus subtilis* 26N led to a significant reduction in OVA-induced food allergy symptoms in BALB/c mice 1 hour after challenge, including diarrhea score ($P < .01$), anaphylactic response score ($P < .01$), and rectal temperature ($P < .01$) [21]. This also lends support to the potential efficacy of these species in preventing food allergy development in murine models.

The gastrointestinal microbiome and food allergy: human studies

Results from human cohort studies support gut dysbiosis in food allergy [6,15,22]. A study by the US-based Consortium for Food Allergy Research (CoFAR) of 141 children from multiple sites across the United States found that children with egg allergy had distinct gut microbiome compositions relative to controls without food allergy (PERMANOVA $R^2 = 2.7\%$, $P = 5.0 \times 10^{-4}$), with Firmicutes and Verrucomicrobia enriched in the gut microbiome of egg allergic children [23]. Logistic regression models adjusted for potential confounders identified the bacterial genera *Ruminococcus*, *Lactococcus*, and *Leuconostoc* as being significantly associated with egg allergy [23]. Independent studies of gut microbiota in subjects with milk allergy have shown similar and contrasting results [6,15,22]. For example, a study of 39 Italian infants found that the gut microbiota of infants with milk allergy had more Lachnospiraceae and Ruminococcaceae [24].

Human cohort studies suggest that gut dysbiosis may precede the development of food allergy. A study of 225 US children from the Vitamin D Antenatal Asthma Reduction Trial (VDAART) found associations between infant gut microbiota and sensitization and food allergy by 3 years of age [25]. Specifically, *Haemophilus*, *Dialister*, *Dorea*, and Clostridium in stool at 3–6 months of age were negatively associated with sensitization to milk, egg, peanut, soy, or wheat; and *Citrobacter*, *Oscillospira*, *Lactococcus*, and *Dorea* in stool at 3–6 months of age were negatively associated with food allergy (based on sensitization and symptoms) by 3 years of age [25]. In an independent study of 56 Japanese infants, the genera *Leuconostoc*, *Weissella*, and *Veillonella* were significantly underrepresented ($P < .05$) in the gut microbiome of infants who later developed food allergies within their first 2 years [26]. A caveat of this study, however, is that food allergy was classified based on self-report and/or questionnaire-based assessments, which can be inaccurate.

Other human cohort studies have demonstrated associations between gut microbial composition and the clinical trajectory of established food allergy [6,15,22]. A study of 226 US infants with milk allergy by CoFAR showed that infant gut microbiota are associated with the future clinical course of established food allergy [27]. Specifically, the Firmicutes phylum and Clostridia class of bacteria were more highly represented in the infant gut microbiomes of subjects 3–6 months of age whose milk allergy later resolved by 8 years of age, whereas Bacteroidetes and *Enterobacter* were more highly represented in milk allergic infants whose milk allergy persisted into mid-childhood [27]. Such findings suggest potential utility in gut microbiome manipulation for therapeutic outcomes.

Microbiome-based therapeutic interventions

Given research thus far demonstrating associations between gut microbiome composition and food allergy, there has also been attention toward modalities to manipulate the gut microbiome for therapeutic intervention [6,15,22,28]. The modalities that have been considered for food allergy prevention and treatment include probiotics (live microorganisms intended to benefit health), prebiotics (substrates that promote the growth of beneficial host microorganisms), synbiotics (combination pre- and probiotics), and fecal microbiota transplantation (FMT, transfer of fecal material from donor to recipient) [6,15,22].

Probiotics

Evidence regarding the usefulness of probiotic treatments for food allergy in humans has been mixed, with some studies reporting a beneficial effect and others reporting no significant benefits. Hol et al. conducted one of the first randomized, controlled trials to explore probiotic supplementation in humans, supplying infants with a combination of *Lactobacillus casei* and *Bifidobacterium lactis* for 12 months [29]. However, the results did not show a significant difference in the cumulative tolerance to cow's milk [29]. Cuello-Garcia et al. performed a systematic review of 2403 articles in the Cochrane Central Register of Controlled Trials published up to December 2014 and concluded that the available evidence was not supportive of the effect of probiotics on food allergies [30]. However, the authors from this systematic review also noted that their certainty in their conclusions was low because of the potential for bias and inconsistency, as well as the use of indirect evidence to explore this question, and that further studies would be needed.

More recently, Lundelin et al. administered four different probiotic strains (*Lactobacillus rhamnosus* GG, *B. lactis* Bb-12, *L. paracasei* ST11, and *Bifidobacterium longum* BL999) perinatally to children [31]. 44% of children receiving *L. rhamnosus* GG perinatally had developed allergic disease, as compared to 56% of children receiving a placebo at 10–15-year follow-up ($P < .047$) [31]. A randomized controlled trial, conducted over 36 months, found that children who were given extensively hydrolyzed casein formula (EHCF) with the probiotic *L. rhamnosus* GG showed a reduction of allergic manifestations, as compared to children given EHCF only at 36 months [32]. Tang et al. administered the probiotic *L. rhamnosus* CGMCC 1.3724, as well as peanut oral immunotherapy in children with peanut allergy; this combination (probiotic and peanut oral immunotherapy [PPOIT]) was associated with reduced peanut skin test responses and peanut-specific IgE levels, as well as an increase in peanut-specific IgG_4 levels [33]. Furthermore, a 4-year follow-up of this trial showed that PPOIT-treated participants were significantly more likely than placebo group participants to have continued eating peanuts in the interval, as well as having smaller reactions to the peanut skin prick test, and significantly higher peanut sIgG4: sIgE ratios [34].

Prebiotics

Studies in mice models have also suggested a protective effect of prebiotics on the development of food allergies. It has been previously suggested that gut microbiota are able to metabolize human milk oligosaccharides [35]. Castillo-Courtade et al. examined the effects of oral treatment of two of these oligosaccharides, 2′-fucosyllactose or 6′-sialyllactose, on attenuating anaphylactic symptoms in OVA-sensitized mice; treatment with either oligosaccharide led to a significant decrease in symptoms upon allergen challenge [36]. Furthermore, it has been shown that short-chain fatty acids (SFCAs) can promote intestinal Treg cells and can inhibit mast cell expansion and degranulation [37,38]. Tan et al. studied the possible protective role of SCFAs against food allergies, by inoculating germ-free mice with zero fiber (ZF) or high (HF) microbiota; anaphylaxis score was significantly decreased in HF versus ZF mice ($P < .05$); when wild-type (WT) and *Myd88*$^{-/-}$ mice were fed on a HF diet and food allergy was induced, the *Myd88*$^{-/-}$ mice had a significantly higher anaphylaxis score ($P < .005$) [39]. Furthermore, mice without GPR43 or GPR109A, receptors for SCFA's, showed exacerbated anaphylaxis scores ($P < .05$ and $P < .01$, respectively compared to WT), even when under conditions of HF diet [39].

There have not been many studies that have examined the effects of prebiotics on food allergies in human subjects. However, Lee-Sarwar et al. found that several metabolites (sphinganine, 3-ketosphinganine, 3-hydroxypalmitate, *N*-palmitoylserine, 13-methylmyristate) that were associated with de novo sphingolipid synthesis were significantly higher in subjects with food sensitization than in those with food allergy ($P = .02$), or controls ($P = .02$), suggesting that these sphingolipids might be able to confer protection against food allergy in individuals who are predisposed to food sensitization [40].

Synbiotics

The provision of prebiotics and probiotics together, termed synbiotics, could be more effective than either alone because the combination provides both desired microbes as well as nutrients to foster their growth [6,15,22]. A Phase 1 study of the synbiotic *Bifidobacterium breve* M-16V plus oligofructose, long-chain inulin, and acidic oligosaccharides added to amino acid-based formula (AAF) versus AAF alone among 110 infants with cow's milk allergy showed normal and similar growth in both arms [41]. A clinical trial is now being performed comparing partially hydrolyzed formula with symbiotic versus standard infant formula (NCT03067714). The primary outcome is fecal levels of *Bifidobacteria* at 17 weeks of age, and secondary outcomes include *Bifidobacteria* levels at 1 year of age and IgE-mediated allergic manifestations, including food allergy, up to 1 year of age (NCT03067714).

Fecal microbiota transplantation

FMT involves transferring communities of microbes from a donor to a recipient. Thus far, FMT has been most notably used for treating *Clostridium difficile* colitis, where fecal material from healthy donors is transplanted to patients with the disease [42,43]. Given murine model findings supporting food allergy-modifying effects from fecal transfer, FMT may be a strategy to consider for food allergy therapy in human beings. A small Phase I open-label trial to evaluate the safety and tolerability of oral encapsulated FMT administered open-label over 2 days for the treatment of peanut allergy in 10 adult subjects is underway (NCT02960074).

Future directions

Studies in murine models as well as observational and interventional human studies have greatly expanded our understanding of the human gastrointestinal microbiome in the context of food allergy. Gut dysbiosis seems to be linked to the pathogenesis of food allergy, among other disorders, and bacterial composition may differ in subjects with and without food allergies. However, there is still much to be explored regarding this relationship. A specific pathway connecting these changes has not yet been elucidated, and it remains to be seen whether the use of probiotics, prebiotics, synbiotics, and FMT in human beings can confer meaningful disease-modifying effects.

Relevance across the life course

Microbiome associations with food allergy have been largely explored in pediatric populations, and research into this relationship in adults has been vastly more limited. In a study of self-reported allergy of all kinds (drug, venom, environmental, and various foods) in adults, Hua et al. reported positive associations between gut microbial richness and the self-reported tree nut, shellfish, and peanut allergy [44]. In this sample, the relative abundance of *Bacteroides* was positively associated with peanut and tree nut allergy at FDR $<10\%$ [44]. A major limitation of this study is that it relied on self-reported allergy, which is notoriously inaccurate. In the absence of other data on microbiome associations with food allergy in adults, it is difficult to state at this point how findings among children may translate across the life course. Further, many food allergies are commonly outgrown by adulthood; for example, 79% of milk allergies resolve by 16 years of age; 68% of egg allergies resolve by 16 years of age; and 21.5% of peanut allergies resolve by 20 years of age [45–47].

References

[1] Sicherer SH, Sampson HA. Food allergy: a review and update on epidemiology, pathogenesis, diagnosis, prevention, and management. J Allergy Clin Immunol 2018;141:41–58.

[2] Bunyavanich S, Rifas-Shiman SL, Platts-Mills TA, Workman L, Sordillo JE, Gillman MW, et al. Peanut allergy prevalence among school-age children in a US cohort not selected for any disease. J Allergy Clin Immunol 2014;134:753–5.

[3] Panel NI-SE, Boyce JA, Assa'ad A, Burks AW, Jones SM, Sampson HA, et al. Guidelines for the diagnosis and management of food allergy in the United States: report of the NIAID-sponsored expert panel. J Allergy Clin Immunol 2010;126:S1–58.

[4] Tordesillas L, Berin MC. Mechanisms of oral tolerance. Clin Rev Allergy Immunol 2018;55:107–17.

[5] Strachan DP. Hay fever, hygiene, and household size. BMJ 1989;299:1259–60.

[6] Bunyavanich S, Berin MC. Food allergy and the microbiome: current understandings and future directions. J Allergy Clin Immunol 2019;144:1468–77.

[7] Koplin JJ, Dharmage SC, Ponsonby AL, Tang ML, Lowe AJ, Gurrin LC, et al. Environmental and demographic risk factors for egg allergy in a population-based study of infants. Allergy. 2012;67:1415–22.

[8] Bager P, Wohlfahrt J, Westergaard T. Caesarean delivery and risk of atopy and allergic disease: meta-analyses. Clin Exp Allergy 2008;38:634–42.

[9] Sepp E, Julge K, Vasar M, Naaber P, Bjorksten B, Mikelsaar M. Intestinal microflora of Estonian and Swedish infants. Acta Paediatr 1997;86:956–61.

[10] Bjorksten B, Naaber P, Sepp E, Mikelsaar M. The intestinal microflora in allergic Estonian and Swedish 2-year-old children. Clin Exp Allergy 1999;29:342–6.

[11] Bunyavanich S. Food allergy: could the gut microbiota hold the key? Nat Rev Gastroenterol Hepatol 2019;16:201–2.

[12] Noval Rivas M, Burton OT, Wise P, Zhang YQ, Hobson SA, Garcia Lloret M, et al. A microbiota signature associated with experimental food allergy promotes allergic sensitization and anaphylaxis. J Allergy Clin Immunol 2013;131:201–12.

[13] Rodriguez B, Prioult G, Hacini-Rachinel F, Moine D, Bruttin A, Ngom-Bru C, et al. Infant gut microbiota is protective against cow's milk allergy in mice despite immature ileal T-cell response. FEMS Microbiol Ecol 2012;79:192–202.

[14] Feehley T, Plunkett CH, Bao R, Choi Hong SM, Culleen E, Belda-Ferre P, et al. Healthy infants harbor intestinal bacteria that protect against food allergy. Nat Med 2019;5:448–53.

[15] Ho HE, Bunyavanich S. Role of the microbiome in food allergy. Curr Allergy Asthma Rep 2018;18:27.

[16] Abdel-Gadir A, Stephen-Victor E, Gerber GK, Noval Rivas M, Wang S, Harb H, et al. Microbiota therapy acts via a regulatory T cell MyD88/RORgammat pathway to suppress food allergy. Nat Med 2019;25:1164–74.

[17] Mortha A, Chudnovskiy A, Hashimoto D, Bogunovic M, Spencer SP, Belkaid Y, et al. Microbiota-dependent crosstalk between macrophages and ILC3 promotes intestinal homeostasis. Science. 2014;343:1249288.

[18] Atarashi K, Tanoue T, Oshima K, Suda W, Nagano Y, Nishikawa H, et al. Treg induction by a rationally selected mixture of Clostridia strains from the human microbiota. Nature. 2013;500:232–6.

[19] Atarashi K, Tanoue T, Shima T, Imaoka A, Kuwahara T, Momose Y, et al. Induction of colonic regulatory T cells by indigenous *Clostridium* species. Science. 2011;331:337–41.

[20] Round JL, Mazmanian SK. Inducible Foxp3 + regulatory T-cell development by a commensal bacterium of the intestinal microbiota. Proc Natl Acad Sci U S A 2010;107:12204–9.

[21] Shin HS, Eom JE, Shin DU, Yeon SH, Lim SI, Lee SY. Preventive effects of a probiotic mixture in an ovalbumin-induced food allergy model. J Microbiol Biotechnol 2018;28:65–76.

[22] Zhao W, Ho HE, Bunyavanich S. The gut microbiome in food allergy. Ann Allergy Asthma Immunol 2019;122:276–82.

[23] Fazlollahi M, Chun Y, Grishin A, Wood RA, Burks AW, Dawson P, et al. Early-life gut microbiome and egg allergy. Allergy. 2018;73:1515–24.

[24] Berni Canani R, Sangwan N, Stefka AT, Nocerino R, Paparo L, Aitoro R, et al. *Lactobacillus rhamnosus* GG-supplemented formula expands butyrate-producing bacterial strains in food allergic infants. ISME J 2016;10:742–50.

[25] Savage JH, Lee-Sarwar KA, Sordillo J, Bunyavanich S, Zhou Y, O'Connor G, et al. A prospective microbiome-wide association study of food sensitization and food allergy in early childhood. Allergy. 2017;73:145–52.

[26] Tanaka M, Korenori Y, Washio M, Kobayashi T, Momoda R, Kiyohara C, et al. Signatures in the gut microbiota of Japanese infants who developed food allergies in early childhood. FEMS Microbiol Ecol 2017;93.

[27] Bunyavanich S, Shen N, Grishin A, Wood R, Burks W, Dawson P, et al. Early-life gut microbiome composition and milk allergy resolution. J Allergy Clin Immunol 2016;138:1122–30.

[28] Ho HE, Bunyavanich S. Microbial adjuncts for food allergen immunotherapy. Curr Allergy Asthma Rep 2019;19:25.

[29] Hol J, van Leer EH, Elink Schuurman BE, de Ruiter LF, Samsom JN, Hop W, et al. The acquisition of tolerance toward cow's milk through probiotic supplementation: a randomized, controlled trial. J Allergy Clin Immunol 2008;121:1448–54.

[30] Cuello-Garcia CA, Brozek JL, Fiocchi A, Pawankar R, Yepes-Nunez JJ, Terracciano L, et al. Probiotics for the prevention of allergy: a systematic review and meta-analysis of randomized controlled trials. J Allergy Clin Immunol 2015;136:952–61.

[31] Lundelin K, Poussa T, Salminen S, Isolauri E. Long-term safety and efficacy of perinatal probiotic intervention: evidence from a follow-up study of four randomized, double-blind, placebo-controlled trials. Pediatr Allergy Immunol 2017;28:170–5.

[32] Berni Canani R, Di Costanzo M, Bedogni G, Amoroso A, Cosenza L, Di Scala C, et al. Extensively hydrolyzed casein formula containing *Lactobacillus rhamnosus* GG reduces the occurrence of other allergic manifestations in children with cow's milk allergy: 3-year randomized controlled trial. J Allergy Clin Immunol 2017;139:1906–13 e1904.

[33] Tang ML, Ponsonby AL, Orsini F, Tey D, Robinson M, Su EL, et al. Administration of a probiotic with peanut oral immunotherapy: a randomized trial. J Allergy Clin Immunol 2015;135:737–44 e738.

[34] Hsiao KC, Ponsonby AL, Axelrad C, Pitkin S, Tang MLK, Team PS. Long-term clinical and immunological effects of probiotic and peanut oral immunotherapy after treatment cessation: 4-year follow-up of a randomised, double-blind, placebo-controlled trial. Lancet Child Adolesc Health 2017;1:97–105.

[35] Koropatkin NM, Cameron EA, Martens EC. How glycan metabolism shapes the human gut microbiota. Nat Rev Microbiol 2012;10:323–35.

[36] Castillo-Courtade L, Han S, Lee S, Mian FM, Buck R, Forsythe P. Attenuation of food allergy symptoms following treatment with human milk oligosaccharides in a mouse model. Allergy. 2015;70:1091–102.

[37] Zhang H, Du M, Yang Q, Zhu MJ. Butyrate suppresses murine mast cell proliferation and cytokine production through inhibiting histone deacetylase. J Nutr Biochem 2016;27:299–306.

[38] Arpaia N, Campbell C, Fan X, Dikiy S, van der Veeken J, deRoos P, et al. Metabolites produced by commensal bacteria promote peripheral regulatory T-cell generation. Nature 2013;504:451–5.

[39] Tan J, McKenzie C, Vuillermin PJ, Goverse G, Vinuesa CG, Mebius RE, et al. Dietary fiber and bacterial SCFA enhance oral tolerance and protect against food allergy through diverse cellular pathways. Cell Rep 2016;15:2809–24.

[40] Lee-Sarwar K, Kelly RS, Lasky-Su J, Moody DB, Mola AR, Cheng TY, et al. Intestinal microbial-derived sphingolipids are inversely associated with childhood food allergy. J Allergy Clin Immunol 2018;142:335–8 e339.

[41] Burks AW, Harthoorn LF, Van Ampting MT, Oude Nijhuis MM, Langford JE, Wopereis H, et al. Synbiotics-supplemented amino acid-based formula supports adequate growth in cow's milk allergic infants. Pediatr Allergy Immunol 2015;26:316–22.

[42] van Nood E, Vrieze A, Nieuwdorp M, Fuentes S, Zoetendal EG, de Vos WM, et al. Duodenal infusion of donor feces for recurrent *Clostridium difficile*. N Engl J Med 2013;368:407–15.

[43] Youngster I, Russell GH, Pindar C, Ziv-Baran T, Sauk J, Hohmann EL. Oral, capsulized, frozen fecal microbiota transplantation for relapsing *Clostridium difficile* infection. JAMA. 2014;312:1772–8.

[44] Hua X, Goedert JJ, Pu A, Yu G, Shi J. Allergy associations with the adult fecal microbiota: analysis of the American Gut Project. EBioMedicine. 2016;3:172–9.

[45] Skripak JM, Matsui EC, Mudd K, Wood RA. The natural history of IgE-mediated cow's milk allergy. J Allergy Clin Immunol 2007;120:1172–7.

[46] Savage JH, Matsui EC, Skripak JM, Wood RA. The natural history of egg allergy. J Allergy Clin Immunol 2007;120:1413–17.

[47] Skolnick HS, Conover-Walker MK, Koerner CB, Sampson HA, Burks W, Wood RA. The natural history of peanut allergy. J Allergy Clin Immunol 2001;107:367–74.

Chapter 8

Microbiome and pediatric obesity, malnutrition, and nutrition

Gregory C. Valentine[1], Amy B. Hair[2] and Camilia R. Martin[3,4]
[1]Division of Neonatology, Seattle Children's Hospital, University of Washington, Seattle, WA, United States, [2]Section of Neonatology, Texas Children's Hospital, Baylor College of Medicine, Houston, TX, United States, [3]Department of Neonatology, Beth Israel Deaconess Medical Center, Boston, MA, United States, [4]Harvard Medical School, Boston, MA, United States

Introduction

Within each individual reside trillions of microorganisms composing the human microbiome. Approximately 3.8×10^{13} bacteria are living among an estimated 0.3×10^{13} nucleated human cells—a 10:1 ratio of bacteria to nucleated human cells [1]. Bacteria composing our human enteric microbiota can serve as commensals, promoting significant health benefits including vitamin and metabolite production and promotion of a healthy intestinal barrier to prevent infection. Conversely, dysbiosis of the enteric microbiome contributes to the pathogenesis of intestinal and systemic diseases. Thus the human microbiome plays an integral role in promoting or impeding a state of healthy living.

The question that has eluded researchers for years is whether perturbations in the early, developing human microbiome lead to an increased susceptibility to develop certain diseases later in life. The developmental origins of health and disease (DOAD) hypothesis postulates that fetal and early neonatal changes can lead to adult-onset metabolic diseases such as cardiac disease [2–5]. Building upon this theory, numerous studies have investigated whether early-life exposures and microbiome alterations are associated with later onset of certain disease states [6–11]. Research has elucidated a potential role between the composition of the human microbiome and the development of common human diseases such as diabetes [12–15], obesity [16–22], metabolic syndrome [18,23], and autoimmune disorders [24]. Thus the initial development of the neonatal microbiome, and any alterations that occur within it as the neonate matures into a toddler, child,

The Developing Microbiome. DOI: https://doi.org/10.1016/B978-0-12-820602-7.00008-8

and then adolescent may have significant effects throughout the individual's lifespan. However, causation has yet to be firmly established.

Similar to the microbiome, nutritional states such as obesity and malnutrition have life-long consequences, and an individual's microbiome and his or her nutritional state are intricately connected [25–27]. Both considered forms of inadequate nutritional intake, obesity and malnutrition, affect millions of individuals globally. For example, in 2015, 108 million children and 604 million adults were obese throughout the world, and the prevalence of obesity has doubled in more than 70 countries since 1980 [28]. A high body mass index (BMI) accounted for 4 million deaths globally [28]. Furthermore, approximately 178 million children under the age of 5 are stunted, and an additional 112 million are underweight [29]. Thus understanding the role of the microbiome and its effects on propagating one's nutritional status could elucidate potential microbiota-directed dietary interventions that could lead to targeted improvements in millions of lives worldwide.

Within this chapter, we will evaluate and review the literature surrounding the impact of maternal diet and maternal obesity on the developing neonatal microbiome, and how breastfeeding or formula feeding practices can have profound effects on microbiome development. Then we will explore the effects of microbiome alterations and the risk of childhood obesity before delving into how childhood malnutrition impacts the pediatric microbiome. Finally, we will conclude this chapter with a look forward into future directions and questions needing evaluation through robust scientific studies.

The impact of maternal diet and maternal obesity on the developing neonatal microbiome

Maternal diet plays an integral role in the development of the neonatal microbiome. Thus we will start our chapter by exploring the literature surrounding how maternal diet, specifically a high-fat diet, alters both the maternal and neonatal microbiomes.

Evaluating the literature surrounding animal models on high-fat diets

The gut microbiome is known to fluctuate based upon the available substrates present in consumed food [30]. Therefore a natural corollary question related to pregnancy is whether ingesting a high-fat diet during pregnancy alters the maternal intestinal microbiome and/or the developing neonatal microbiome. Ma et al. derived an eloquent primate model to address this specific question [31]. *Macaca fuscata* (Japanese macaque) pregnant dams were fed either a high-fat or control diet, both of which were isocaloric, and continued on this diet over the course of gestation and through the period of lactation. The offspring were weaned from breastfeeding at approximately 6–7 months of age and then either continued on the same diet or switched to the other diet formulation. The pregnant dams consuming a high-fat diet developed significant changes in the maternal intestinal microbiome compared to controls, namely increased Bacteroidetes and decreased Spirochetes. Interestingly, a high-fat maternal diet was associated with persistent changes in the neonatal microbiome that persisted through 1 year of age (such as a decrease in the commensal species *Campylobacter*), and did not fully correct if the infant primate was switched to a control, low-fat, isocaloric diet. While partial correction did occur, persistent differences remained, and these results showcase the importance of the maternal diet during gestation and breastfeeding on the developing neonatal and infant enteric microbiome [31]. This discovery further promotes the DOAD hypothesis, which theorizes an early period of unique susceptibility to metabolic programming that occurs in the fetal and early neonatal time period. Furthermore, these findings suggest that rather than the microbiome leading to obesity or overweight body habitus, the diet was integral in altering both the maternal and the developing neonatal enteric microbiomes.

Showcasing the importance of the maternal diet, the same researchers found Japanese macaque offspring exposed to a high-fat maternal diet during pregnancy also had diminished thyroid hormone production, altered hepatic circadian gene expression (*Npas2*), increased anxiety behaviors, and a higher likelihood of developing nonalcoholic fatty liver disease [11,32–35]. Notably, while hepatic circadian gene expression was altered by a maternal high-fat diet, the disruption was fully corrected if the high-fat diet was discontinued and a control diet was commenced postnatally [10]. Thus high-fat diet, even if isocaloric and in nonobese hosts, leads to dysbiosis within the enteric microbiome. Diet, rather than mere obesity, plays an integral role in altering the developing enteric microbiome, and these findings highlight potential targets for therapeutic remedies such as dietary changes during the peripartum period.

An assessment of human-based studies on maternal diet and the maternal and developing neonatal microbiome

While animal models help elucidate potential mechanisms, the findings must correlate to what occurs in humans. We now turn our attention to human-based studies and evaluate the literature by evaluating the maternal and developing neonatal microbiome.

The maternal microbiome is known to fluctuate based upon obesity, overweight status, and gestational diabetes mellitus [36,37]. Gomez-Arango et al. investigated whether there were alterations within the maternal gut microbiome and associated changes in metabolic hormones in obese or overweight pregnant women at 16 weeks gestation. Interestingly, the growth of specific bacteria was promoted or diminished in association with specific hormones. Positive correlations were found between gastrointestinal polypeptide and *Coprococcus* (butyrate producers, an important short-chain fatty acid (SCFA) that can serve as a source of energy), insulin and *Collinsella* (colonizes mucosal surfaces and helps metabolize amino acids), and Adipokine and *Ruminococcaceae* and *Lachnospiraceae* (both help metabolize indigestible polysaccharides into SCFAs). Negative correlations existed between gastrointestinal polypeptide and *Ruminococcaceae* [37]. In addition, overweight women had higher microbial richness than obese women [37]. This landmark study elucidated a potential relationship between metabolic hormones and the microbiome. Hormones linked to weight gain were associated with bacteria that help produce energy via metabolism of substrates and production of SCFAs, potentially promoting obesity further. At a minimum, these results showcase how the gut microbiome can influence the metabolism of the human host. However, it remains to be determined whether the specific bacteria predispose to the development of certain chronic conditions such as obesity, or the hormonal changes drive the changes within the enteric microbiome. Further investigations are also warranted in understanding how these gestational microbiome alterations associated with chronic diseases affect the developing fetal, neonatal, and infant microbiomes.

To investigate how a maternal high-fat diet affects the developing neonatal microbiome, Chu et al. evaluated 81 neonates by sampling meconium and multiple body sites at delivery and at 6 weeks of age [38]. Mothers underwent a dietary questionnaire and were classified as either a control group or high-fat group based on their average intake of fat. Notably, neonates who were born to women with a high-fat diet had significantly lower proportions of *Bacteroides* when compared to the control group. This effect persisted from birth through 6 weeks of age.

These data support that a maternal high-fat diet alters not only the maternal enteric microbiome, but also the developing neonatal microbiome highlighting the important role of micro- and macronutrient intake during pregnancy and how dietary changes can have profound and lasting impacts on the evolving neonatal microbiome.

Breastfeeding versus formula feeding and the neonatal microbiome

The American Academy of Pediatrics (AAP) recommends exclusive breastfeeding for the first 6 months of life and up to 1 year or longer due to the significant health benefits associated with breastfeeding compared to formula feeding [39]. For instance, neonates who exclusively breastfeed and/or receive mother's own milk have decreased rates of otitis media [40], gastrointestinal tract infections [40–43], necrotizing enterocolitis (NEC) [40,44], sudden infant death syndrome [40,45–48], and even infant mortality [49].

The mechanism by which breastfeeding and breast milk provide such significant health benefits to the developing neonate are multifold. Breast milk is more than just a source of nutrition; it contains fatty acids, vitamins, minerals, secretory immunoglobulin A (IgA), and even its own microbiome [50]. Secretory IgA provides passive immunity to the neonate and protects breastfeeding neonates from infections. In addition, "bifidus factor" was first found in human breast milk in 1926 by Schonfeld et al. [51,52] and is so named because it was discovered to promote the growth of *Bifidobacterium bifidus*, a protective commensal species. "Bifidus factor" is now more commonly known as human milk oligosaccharides (HMOs).

Human milk oligosaccharides

HMOs are the third most common component of breast milk after fat and carbohydrates, and they foster the growth of healthy, commensal bacteria such as *Bifidobacterium infantis* [53]. The promotion of commensal growth effectively prevents pathogenic bacteria from taking residence in the developing neonatal microbiome. In addition, HMOs prevent the attachment of bacteria, parasites, and other viruses to the intestinal wall, thus promoting a more effective intestinal barrier to prevent neonatal and infant infection. Studies have shown that HMOs help prevent the attachment of human immunodeficiency virus (HIV) to attachment receptors within the intestine [54,55]. This feature of HMOs may be why vertical HIV transmission rates in exclusive breastfeeding are less than 10%.

Because of the health benefits attributed to HMOs, formula companies are now considering adding HMOs as supplements to their formula products. Further research is needed to explore the impact of supplemental HMOs to formula and whether they provide the same benefit as seen in breast milk. Regardless, while there are hundreds of specific HMOs in breast milk, studies are currently exploring adding only 1–2 HMOs as supplements to their formula products. Questions remain, however, whether selecting a few HMOs from a complex matrix of many would attenuate any potential benefit of supplementation.

The breast milk microbiome

While breast milk was once considered to be sterile, we now know that it contains its own microbiome [56,57]. While this is an area currently under significant evaluation via numerous active research studies, it is known that breast milk contains key bacteria that are generally regarded as common to the skin (*Staphylococcus*), oral cavity (*Streptococcus*), and intestinal flora (*Bifidobacteria*, *Enterococcus*, and *Escherichia*) [56]. When a neonate breastfeeds, the breast milk microbiome seeds the neonatal gut, likely aiding in the development of healthy commensals. How the breast milk microbiome is established is under investigation but is believed to be due to a process called enteromammary trafficking. This process postulates that bacteria in the maternal gastrointestinal tract are engulfed by phagocytic cells such as leukocytes and then translocated to the mammary glands via the system circulation [58] and is likely the reason why one study found 60% of women given oral *Lactobacillus* probiotics had *Lactobacillus* found in their breast milk after probiotic administration [59]. Thus these gut-derived bacteria likely play a key role in the development of the neonatal and infant gut microbiome. In addition, given the significant association of maternal diet on the maternal enteric microbiome, it is highly likely the breast milk microbiome differs based on diet. Further research is necessary to evaluate the potential effect of a high-fat maternal diet on the breast milk microbiome, its association with the developing neonatal microbiome, and potential later development of childhood obesity.

Maternal secretory immunoglobulin A

Another component of breast milk that provides protection against disease development in neonates is secretory immunoglobulin A (IgA), an antibody derived from the mother, which provides passive immunity to the breastfeeding neonate. Research has shown that rates of NEC are significantly lower in neonates consuming maternal milk as opposed to formula [60]. However, the underlying protective mechanisms were previously unknown. As aforementioned, HMOs and the breast milk microbiome likely lead to this protection, but a recent publication by Gopalakrishna et al. elucidated the role of secretory IgA as a paramount contributor. When newborn mice were nurtured by IgA-deficient mothers, they had a higher rate of NEC even after exposure to maternal milk [61]. Thus maternal secretory IgA is a beneficial agent within human milk that promotes health in premature neonates.

In conclusion, breast milk provides numerous health benefits to the maturing neonate including fostering the development of healthy commensal species via its own breast milk microbiome, HMOs, secretory IgA, and likely numerous other factors. Exploring the impact of maternal diet on the evolving breast milk microbiome, the role of enterohepatic trafficking in situations of high-fat diet or malnutrition states, and the impact on the developing neonatal microbiome are significant areas in need of further research to fully understand the critical role of breast milk on the developing neonatal microbiome.

Association of microbiome alterations and risk of childhood obesity

Childhood obesity has now become one of the most significant issues facing youth in developed countries throughout the world, especially in the United States, and it is now facing epidemic proportions. In fact, among children aged 2–5 years of age, rates of overweight and class I obesity significantly increased from 2001 to 2015: rates of children classified as overweight increased from 21.2% to 26% and class I obesity increased from 10.7% to 13.7% [62]. When looking at all ages and sexes from 2 to 19 years of age, the overall rate of being overweight in 2015 was 35.1%, and class I obesity was 18.5% [62]. Thus more than 1 out of every 3 children is overweight and nearly 1 out of every 5 children is classified as obese. Moreover, children who are obese are known to have a greater risk of adverse health outcomes, comorbidities, and premature death [63]. Reflective of the seriousness of the obesity epidemic and implications on adult health, the AAP now advocates for pediatricians to consider bariatric surgery for children with morbid obesity [64].

While obesity is multifactorial in origin, what relationship does the microbiome have with the development of obesity? Do perturbations in the developing microbiome during the uniquely susceptible time period of fetal or neonatal life affect the later development of obesity? Here, we seek to explore the answers to these paramount and pressing clinically important questions. With clarity of these questions, it is possible that altering the microbiome in children at risk for obesity may be a potentially more advantageous option versus bariatric surgery imposing less of an invasive risk to the patient.

Earlier in this chapter, we discussed the studies that elucidated the role of maternal diet, specifically a high-fat diet, and its impact on both the maternal and developing neonatal enteric microbiomes. These studies implicated diet rather than an obesogenic microbiome as the instigator to obesity or overweight status. However, the aforementioned study by Gomez-Arango et al. eloquently demonstrated a clear relationship between the enteric microbiome and specific hormone levels [37] raising the question of whether changes in the microbiome itself can lead to predispositions in the development of childhood obesity.

What do animal models tell us?

Leptin is a hormone secreted by adipocytes that is known to inhibit feelings of hunger. Therefore leptin-deficient mice have become a well-regarded obese mouse model. Utilizing this animal model, Ley et al. discovered decreased abundance of Bacteroidetes and higher abundance of Firmicutes in the enteric microbiome when compared to lean mice [65]. These findings of higher Firmicutes and lower Bacteroidetes associated with obesity have been reproduced in several other animal models produced from the same group of researchers [20,21,66]. Other groups of researchers have also found significant alterations within the microbiomes associated with an obese phenotype when compared to their lean counterparts. For example, Waldram et al. utilized a rat-based obesity animal model and found higher abundance of Firmicutes

(*Clostridium* and *Lactobacillus* sp.) and decreased abundance of Actinobacteria (*Bifidobacterium* and *Atopobium* sp.) within the enteric microbiome [67].

Furthermore, in an eloquently designed study, Murphy et al. evaluated the gut microbiome at three different time points in genetically determined obese mice (*ob/ob*) who were fed a low-fat diet, and wild-type mice who were either fed a low-fat diet or a high-fat diet for 8 weeks [68]. Notably, a progressive increase in the abundance of Firmicutes occurred in the high-fat diet fed wild-type mice and the *ob/ob* mice, and Firmicutes remained stable in the low-fat diet fed wild-type mice. When the investigators evaluated for changes in energy metabolism markers, changes in the microbiome did not correlate with changes in these markers [68]. Therefore this study suggested that it was more likely that diet rather than the underlying obesity led to the changes in the enteric microbiome.

Animal models have elucidated an association between obesity, diet, and the enteric microbiome. Obesity is associated with higher abundance of certain bacteria (Firmicutes) and lower abundance of others (Bacteroidetes). In addition, a high-fat diet in wild-type mice and genetically modified obese mice had progressively increased abundance of Firmicutes compared to their lean counterparts. Thus in animal models, a clear difference is evident between the enteric microbiomes of obese and lean individuals. What appears to emerge through the amassing medical literature is that diet remains the primary instigator on the resultant microbiome, which then metabolically propagates the phenotype of a malnourished, obese, or healthy phenotype. A positive biologic feedback loop is established in which the altered developing microbiome then leads to persistence of the host's nutritional status that initially led to the altered microbiome.

How do these findings correlate with human studies?

Consistent with the aforementioned animal models, when Ley et al. evaluated the enteric microbiome of 12 obese human subjects, they found higher abundance of Firmicute*s* and lower Bacteroidetes in obese subjects compared to their lean controls [69]. Armougom et al. reproduced these findings when they evaluated the enteric microbiome of 20 obese individuals, 9 with the diagnosis of anorexia nervosa, and 20 normal weight subjects considered to be healthy controls. These researchers discovered significantly higher proportions of Firmicutes (*Lactobacillus* sp.) in obese subjects when compared to the lean controls or the subjects with anorexia nervosa. Interestingly, subjects with anorexia nervosa had a higher abundance of *Methanobrevibacter smithii* than their lean counterparts. Thus these findings likely demonstrate how the enteric microbiome changes based on nutritional substrates consumed by the human host and confirm the findings of higher Firmicutes and lower Bacteroidetes seen in animal studies of obesity.

Twins provide a scientifically robust manner in which to evaluate changes in the microbiome associated with environment and genetics. While controlling for the environment, twins offer insights into the role of genetics as monozygotic (MZ) twins share the same genetics compared to dizygotic (DZ) twins which are nonidentical genetically. Turnbaugh et al. assessed 31 MZ and 23 DZ twin pairs who were all young adults (25–32 years old) and generally concordant for obesity or leanness [22]. DZ twins had the same degree of similarity in the gut microbiotas when compared to those of MZ twin pairs supporting the powerful influence of the environment over genetics, although it is important to note that these samples were taken as adults and not in the early, developing neonatal microbiome. Moreover, lower proportions of Bacteroidetes and higher abundance of Actinobacteria were present in obese subjects when compared to lean individuals, but no differences were present in Firmicutes which differ from findings of other studies. Obesity was also associated with decreased microbial diversity [22]. Further supporting a lower level of Bacteroidetes in obesity, another study discovered a decrease in abundance of Bacteroidetes in obese humans that actually increases with weight loss when subjects were changed to a low-calorie diet [69]. Although, these studies were conducted in young adults, similar implications have been found in young children and infants.

Two scientifically robust studies evaluated if specific bacteria present in the enteric microbiome during early postnatal microbiome maturation are associated with later development of obesity. First, the large KOALA Dutch study prospectively evaluated healthy pregnant women who were enrolled at 34 weeks gestation [70]. Of this cohort, 909 one-month-old infants' enteric microbiomes were evaluated for any associations between the abundance of certain bacteria and the later development of obesity. The study found that colonization with *Bacteroides fragilis*, a member of phylum Bacteroidetes, was associated with higher body mass index (BMI) z-score between 1 and 10 years of age among infants with a low fiber intake at age 4. However, of those that were colonized, a higher count of *B. fragilis* was associated with a lower BMI z-score after adjustment for potential confounders [70]. Second, Vael et al. evaluated the enteric microbiome of 138 infants at 3, 26, and 52 weeks of age and evaluated the association of certain bacteria with the later onset of childhood obesity [71]. Higher abundance of *B. fragilis* and lower abundance of *Staphylococcus* within the enteric microbiome of infants between 3 weeks and 1 year of age was associated with a higher risk of obesity during the first 3 years of life [71]. Thus early colonization during the neonatal period with Bacteroidetes appears to be a marker for increased risk of developing childhood obesity.

Evaluating the enteric microbiomes of obese and overweight children, Bervoets et al. prospectively evaluated the enteric microbiomes of 26 overweight/obese and 27 lean children aged 6–16 years old [72]. Obese and overweight children had higher abundance of Firmicutes such as *Lactobacillus* spp., and there was no significant difference in rates of *B. fragilis* between obese and lean children [72]. In addition, *Lactobacillus* was associated with higher

C-reactive protein (CRP) concentrations, a marker of systemic inflammation, in obese children [72]. In fact, *Lactobacillus* spp. were in higher concentrations within the fecal microbiome of obese children compared to healthy controls [72]. While Bervoets et al. did not specify which specific species of *Lactobacillus* were of higher prevalence in obese individuals, Million et al. demonstrated that *Lactobacillus reuteri* was associated with obesity in adults [73]. Consistent with these findings, Ismail et al. found an association between higher plasma CRP and increased levels of Firmicutes in obese children and adults [74]. However, the question remains whether a specific species of *Lactobacillus* is the culprit behind the increased pro-inflammatory signaling cascade, which is common physiologic process of obesity. Nevertheless, there does appear to be a clear association between higher prevalence of *Lactobacillus* and low-grade systemic inflammation. Hence, while early colonization with Bacteroidetes appears to be associated with later development of childhood obesity, higher abundance of Firmicutes and its associated low-grade inflammation emerges as associated with older children and teenagers suffering from obesity. Confirming the association of the enteric microbiome with growth patterns and inflammation, Lu et al. demonstrated that when enteric microbiomes of preterm infants with adequate and poor growth were transfaunated into germ-free mice, the growth patterns were recapitulated with elevations of inflammatory markers such as NF-κB in the mice receiving the microbiome from a preterm infant with poor postnatal growth [75]. All in all, these studies show that there appear to be clear associations between the specific derangements within the enteric microbiome, the promotion of inflammation, and regulation of growth and weight gain. Thus the natural next question becomes: can modification of intestinal microbiota prevent the later development of obesity?

Antibiotic exposure in infancy and association with development of childhood obesity

One of the easiest ways to modify the enteric microbiome is through the administration of oral or intravenous antibiotics. Thus researchers have explored the effects of antibiotics in infancy and its association with later development of obesity. Evaluating 64,580 children for antibiotic exposure until 24 months of age and development of obesity in early childhood (24–59 months of age), Bailey et al. discovered that early exposure to broad-spectrum antibiotics was associated with later development of childhood obesity [76]. Furthermore, repeated exposure to antibiotics was associated with early childhood obesity. Narrow-spectrum antibiotics were not associated with childhood obesity [76]. The mechanism leading to these effects is unclear. However, it is postulated that antibiotic exposure leads to reduced diversity within the enteric microbiota, a lack of colonization with beneficial bacteria, and a resulting inflammation-inducing dysbiosis [77]. Supporting this hypothesis, studies have shown that even one dose of antibiotics early in life can lead to dysbiosis of the enteric microbiome that can last for years [78].

Adding to the evidence of early-life exposure to antibiotics and later development of obesity, Azad et al. studied antibiotic exposure within the first year of life and central adiposity and overweight measurements at 9 and 12 years of age [79]. They found that infants exposed to antibiotics were more likely to be overweight in childhood when compared to infants never exposed to antibiotics, but this effect was present only in boys after adjusting for potential confounding variables [79]. Moreover, Ajslev et al. evaluated 28,354 mother–child dyads from the Danish National Birth Cohort, and found that antibiotic exposure within the first 6 months of life was associated with an increased risk of being overweight among children who had mothers who were not overweight or obese [80]. Interestingly early-life antibiotics were found to confer a decreased risk of being overweight among children whose mothers were either overweight or obese [80]. The authors postulated that the different microbiomes of lean versus overweight mothers were passed on to their infants during parturition, and antibiotics during early infancy can modify the microbial composition either positively or negatively. This theory could explain why infants born to lean mothers had higher rates of obesity after exposure to antibiotics, and infants born to overweight mothers had lower rates of obesity after exposure to antibiotics.

Modulation of the enteric microbiome through the use of broad-spectrum antibiotics is associated consistently within the medical literature with the later development of childhood obesity. Yet, narrow-spectrum antibiotics do not appear to confer this association. Taken in context with the previously mentioned evidence supporting association with certain bacteria such as *Bacteroides* with obesity, modulation of the developing neonatal enteric microbiome during this unique early time period of infancy appears to correlate with the DOAD hypothesis.

Childhood malnutrition and impact on the pediatric microbiome

While obesity is a significant issue affecting developed regions of the world, malnutrition is a pressing issue affecting millions of children globally [81]. While resource-limited settings are commonly challenged with malnutrition, there exist numerous regions throughout the United States that have food insecurity, a major risk factor for the development of malnutrition. In addition, malnutrition during childhood has lasting effects that include growth failure, impaired immunity, recurrent infections, and poor neurodevelopment, and overall mortality [82–84]. While we have discussed the impact of obesity and high-fat diets on the enteric microbiome, we now focus on the opposite end of the nutrition spectrum—exploring the impact of malnutrition and persistent associated perturbations within the enteric microbiome.

The microbiome differs based on geography and diet

Intake of specific dietary factors plays an integral role in the development and modification of the enteric microbiome. A group of researchers evaluated potential differences in the intestinal microbiomes of children aged 1–6 years old from a rural African village in Burkina Fasso compared to those living in Italy [85]. The Burkino Fasso diet was considered to be low in fat and animal protein but high in starch and fiber, whereas the Italian diet was considered to be high in fat, animal protein, and starch but low in fiber. Notably, children from Burkino Fasso had higher levels of Bacteroidetes and lower levels of Firmicutes along with increased quantities of SCFAs. *Prevotella* and *Xylanibacter*, bacteria known to hydrolyze cellulose and xylan, were both found almost exclusively within the enteric microbiome of children from Burkino Fasso suggesting that these bacteria enter a symbiotic relationship as they metabolize the high fiber and polysaccharide diet within Burkino Fasso [85]. Other researchers have also evaluated and shown specific microbiomes associated with resource-limited settings and varying geographical regions [86–88]. Furthermore, Gehrig et al. collected fecal samples from Bangladeshi children with severe acute malnutrition (SAM) and then transfaunated the microbiota into germ-free mice. These mice were then fed different microbiota-directed complementary foods and found alterations in biomarkers, growth mediators, neurodevelopment, and immune function to those of more healthy individuals [89]. Thus the enteric microbiome changes based on the nutritional substrates present in the dietary intake, and microbiota-directed food intake is a promising area of exploration into potential therapeutic dietary interventions that promote a healthy enteric microbiota even in settings of malnutrition.

Gehrig et al. established that the enteric microbiome in children who were suffering from SAM possessed a microbiome pattern that was more immature in its usual chronologic evolution compared to children of the same chronological age who were not malnourished [89]. This finding is consistent with other studies which also found the association of a younger enteric microbiome with malnutrition; and an immature enteric microbiome was related to diminished rates of lean body mass gain, metabolic abnormalities, and perturbations in bone growth [90]. Children suffering from SAM had more immaturity of their enteric microbiome than children suffering from moderate acute malnutrition (MAM), and the effects of SAM were persistent even with alterations in the subject's dietary intake [89]. However, when children with MAM consumed supplemental foods designed to enhance the enteric microbiota, they had improvements in mediators of growth, bone formation, immune function, and neurodevelopment [89].

Thus nutritional intake is integrally related to the development of the enteric microbiome, and a lack of nutritional intake such as in settings of malnutrition is directly associated with an immature enteric microbiota. This immature enteric microbiota further propagates poor nutritional status of the host by altering growth mediators. However, by targeting the enteric microbiota through microbiota-directed foods, these negative effects are partially if not almost completely ameliorated showcasing the ever intertwined and interconnected relationship between one's nutritional status and the enteric microbiome.

The microbiome and acute malnutrition

Malnutrition occurs in stages. First, one develops signs of acute malnutrition such as weight loss and micro- and macronutrient deficiencies. After a longstanding period of time, chronic malnutrition occurs leading to stunting of growth. Here, we delve into the associations of variations within the enteric microbiome associated with acute malnutrition.

One of the most extreme forms of malnutrition is kwashiorkor, which manifests as edema and ascites due to a diet lacking adequate protein intake. Reflecting the association of kwashiorkor with malnutrition, the very name kwashiorkor originates from Ghana and translates to mean, "the sickness the baby gets when the new baby comes." A novel twin study based in Malawi elucidated a further link between malnutrition and the enteric microbiome by evaluating 317 Malawian MZ and DZ twin pairs throughout the first 3 years of life. During the study period, 43% of the twins became discordant and 7% became concordant in developing acute malnutrition [91]. Remarkably, there was no difference in incidence of malnutrition discordance between MZ versus DZ twins. Fecal microbiota samples were obtained from three discordant twin pairs with kwashiorkor and then transplanted into adult 8-week-old germ-free mice, after which the mice were fed staple foods from rural Malawi. Mice that received the kwashiorkor twin fecal transplants developed significantly more weight loss than mice receiving the fecal transplant of the normal weight twin (Fig. 8.1). However, the weight loss was only evident when mice consumed the Malawi diet but not if they consumed standard mouse chow. Bacteria that were of higher abundance in mice with kwashiorkor were *Bilophila wadsworthia*, associated with inflammatory bowel disease, and *Clostridium innocuum*, associated with infections in immunocompromised individuals.

Furthermore, when examining the urinary excretion of energy metabolites from the tricarboxylic acid (TCA) cycle, mice with the kwashiorkor microbiota exhibited signs of selective enzymatic inhibition of the TCA cycle, specifically inhibition of succinate dehydrogenase, whereas mice containing the healthy twin's microbiota did not. Thus the kwashiorkor microbiota may produce inhibitors of energy-production pathways such as TCA cycle enzymes which may make energy metabolism and nutrient regulation more difficult when exposed to a low calorie, poorly nutritious diet. This eloquently performed study

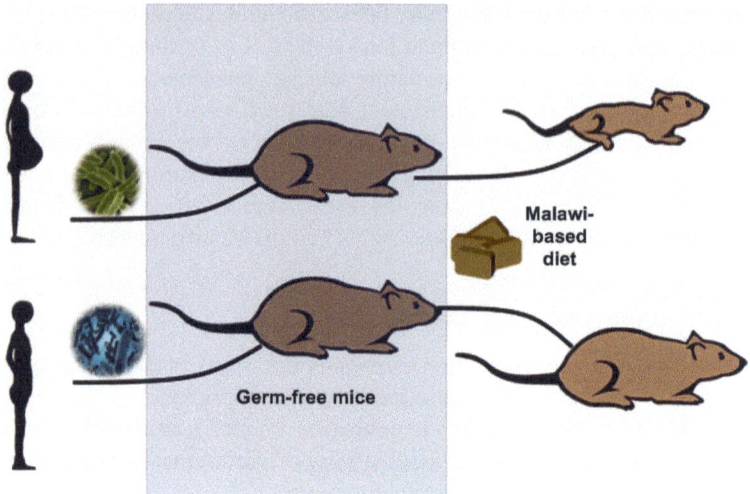

FIGURE 8.1 Smith et al. evaluated the enteric microbiome of pediatric twins in Malawi who were discordant for the development of kwashiorkor versus a healthy weight. The enteric microbiome was transfaunated into 8-week-old germ-free mice, and, interestingly, mice receiving a Malawi-based diet developed weight loss only in the kwashiorkor-microbiota group. These findings showcase a direct relationship between the enteric microbiome and the development of a specific phenotype, thus elucidating a potential biologic feedback loop. *Based on Smith MI, et al. Gut microbiomes of Malawian twin pairs discordant for kwashiorkor. Science. 2013. Available from: https://doi.org/10.1126/science.1229000.*

elucidates a connection between a calorie and nutrient-deficient diet, malnutrition, and the enteric microbiome. Even after transplanting the microbiome into healthy mice, a specific phenotype of malnutrition and weight loss becomes evident in the mice receiving a malnourished microbiome. These mice also exhibit signs of impairment of energy metabolism likely secondary to the microbiome itself—thereby showcasing a potential direct correlation between dysbiosis of the microbiome and the development of malnutrition, specifically kwashiorkor. Moreover, the microbiome in states of malnutrition may impede the ability of the individual to gain weight with proper nutrition given the inhibition of energy metabolism pathways.

The microbiome and chronic malnutrition

Chronic malnutrition results after a prolonged period of micronutrient deficiency. Also known as stunting, chronic malnutrition affects millions of children globally. Forty-three percent of children under 5 years of age are underweight in South and South East Asia [92], and stunting is known to affect an estimated 165 million children in 2011 [81]. How does the microbiome change in settings of chronic malnutrition?

In order to better understand the microbiome within stunted children, Vonaesch et al. evaluated the duodenal, gastric, and fecal specimens from both stunted and nonstunted children aged 2–5 years old living in Bangui, Central African Republic, and Antananarivo, Madagascar [93]. Notably, small intestinal bacterial overgrowth (SIBO), specifically with bacteria known to reside within the oral cavity, was found within the microbiomes of children suffering from stunting. In addition, *Escherichia coli*, *Shigella*, and *Campylobacter* species, known to be pathogens, were found in increased abundance within the enteric microbiomes of children with stunting [93]. Thus small intestinal bowel overgrowth, a form of dysbiosis of the enteric microbiome, and increased prevalence of pathogenic organisms are present in the gastrointestinal tract of children suffering from chronic malnutrition.

Further evidence supporting the role of SIBO contributing to stunting is provided by Donowitz et al. Evaluating 90 two-year-old children living in Bangladesh, the researchers found 16.7% of children had SIBO. Stunting was the strongest predictor for having SIBO as well as having an open sewer outside the home of the child. Notably, increased markers of intestinal inflammation, fecal calprotectin and fecal Reg 1β, were elevated in SIBO positive children when compared to SIBO negative children. Hence, SIBO is associated with stunting and increased signs of intestinal inflammation [94].

Small intestinal bacterial overgrowth and future directions

SIBO adversely impacts the human host in several ways [95]. First, bacteria can deconjugate bile acids that are needed for the absorption of lipophilic vitamins (A, D, E, and K) [96]. Furthermore, bacteria compete for nutrients leading to host micro- and macronutrient deficiencies, decreased SCFA production, increased enteric and systemic inflammation, and potentially increased intestinal permeability [93,94,97–99]. Normally, the human intestinal microbiome performs critical functions in the absorption, metabolism, and even production of vitamins and minerals. In SIBO, dysbiosis leads to a dysregulation of these processes. Thus SIBO, which has a predilection for malnourished individuals, leads to bacterial-host competition for nutrients with decreased absorption of fat-soluble vitamins, increased gut inflammation, and increased intestinal permeability compounding the nutritional status of an already malnourished host.

Clinically, SIBO does not have a uniform presentation of signs or symptoms. Patients can suffer from bloating, flatulence, abdominal pain, diarrhea, steatorrhea, weight loss, or even just nausea and fatigue [100–103]. Some patients may be more symptomatic than others. Thus it is oftentimes overlooked as a diagnosis given the vague nature and potential paucity of symptoms in patients suffering from SIBO.

For decades, antibiotics have been the mainstay therapy for SIBO. A meta-analysis found that antibiotics were superior to treating SIBO over placebo [104], and another systematic review found that treatment with rifaximin led to an overall eradication rate of 70% in patients with SIBO [105]. However, these studies were conducted in adults and not children. Investigators have evaluated the use of trimethoprim–sulfamethoxazole and metronidazole twice daily for 2 weeks in children with SIBO living in poor socioeconomic areas, and there was a 95% eradication rate with this antibiotic regimen [106].

Probiotics are another area of exploration in the potential treatment of SIBO. When compared to placebo, probiotics are associated with increased eradication of SIBO. Remarkably, probiotics outperformed metronidazole in the eradication of SIBO in adults [107]. However, while SIBO appears intricately connected to stunting, neither probiotics nor targeted antibiotic approaches to treatment of SIBO have been thoroughly explored as a potential treatment to acute or chronic malnutrition in resource-limited settings.

Implications of early dysbiosis of the enteric microbiome, obesity, and malnutrition in adult health

Overall, there exists an intricately connected and intertwined relationship between the enteric microbiome and the nutritional status of the human host (Fig. 8.2). Alterations in the diet lead to variations within the microbiota, and dysbiosis of the microbiota can lead to persistent inflammation and the promotion of deficient nutritional health states. Perturbations of the enteric

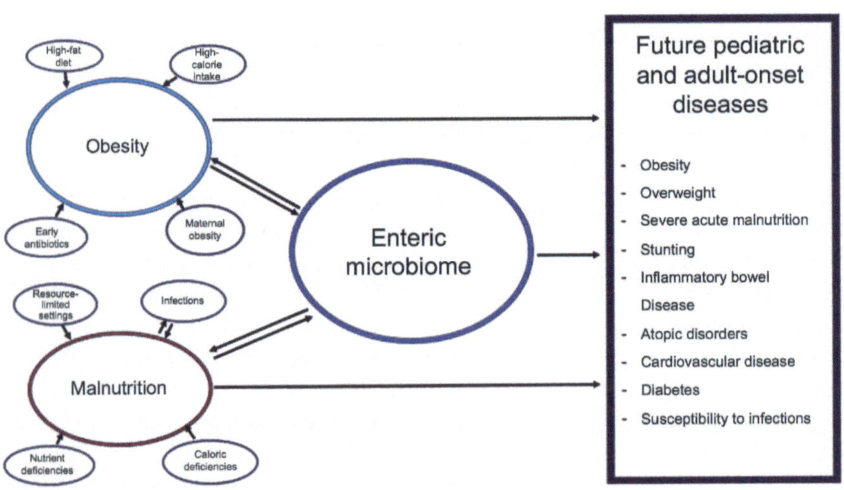

FIGURE 8.2 Individuals suffering from malnutrition or obesity have specific alterations within their enteric microbiomes, and these microbiomes lead to metabolic changes within the human host, which propagate the phenotype of obesity or malnutrition. Thus the intricately interconnected enteric microbiome and malnutrition or obesity leads to long-term adult-onset diseases such as diabetes and cardiovascular disease. Consistent with the developmental origins of health and disease hypothesis, alterations within the enteric microbiome early in life have significant and lasting effects on an infant's life potentially decades later.

microbiome within the critically important developmental window of the neonatal and infant time periods can have lasting effects for years, as seen with antibiotic exposure and later development of childhood obesity. Furthermore, dysbiosis of the enteric microbiome is associated with chronic malnutrition and stunting. Establishing a "healthy" microbiome during this critical developmental time period appears to be integral to the future health and well-being of the human host. A child with malnutrition has a dysregulated enteric microbiome which then effectively alters energy metabolism leading to a positive reinforcement pathway in which malnutrition is then further propagated. Furthermore, there exist specific microbiota alterations associated with later development of childhood obesity, and it is well-known that children with obesity are at a significantly higher risk of having obesity as adults with the accompanying diseases such as cardiovascular disease and diabetes. Thus targeted interventions to promote the development of a "healthy" microbiome during the neonatal and early infantile time period are critical for potential prevention of later development of noncommunicable diseases.

Future directions

While much has become clear over the past few decades regarding the clear association of alterations within the human microbiome and human health and disease, many pressing questions still remain unanswered.

Further exploration into the breast milk microbiome is necessary to classify how maternal dietary changes affect the bacterial members constituting the breast milk microbiome. How do these potential changes affect the colonization of the developing neonatal microbiome in breastfed neonates? Are there specific bacteria present within the breast milk microbiome that prevent or predispose breastfed neonates to the later development of childhood nutritional diseases such as obesity or malnutrition? These are some of the significant questions remaining unanswered that need targeted research studies to help seek out clear answers. We must first understand the answers to these questions in order to then determine potential interventions to target the prevention of childhood nutritional disorders.

Related to childhood obesity, a number of questions remain. While lower abundance of Bacteroidetes and higher proportions of Firmicutes within the enteric microbiome are associated with childhood obesity, do targeted changes of the ratio of Bacteroidetes to Firmicutes change the phenotype of the child? Moreover, while broad-spectrum antibiotic use during infancy is associated with later development of childhood obesity, are there specific bacteria that are reduced by antibiotics that may be ameliorated via therapies such as pre or probiotics? In addition, do antibiotics have the same effects on the premature babies and term babies' microbiome? These and further questions need exploration to help elucidate a deeper understanding of how to potentially prevent and treat childhood obesity by targeting the enteric microbiome.

Finally, SIBO appears strongly associated with chronic malnutrition and stunting. As such, do children living in resource-limited settings need to take prophylactic probiotics in order to prevent the development of severe malnutrition by promoting a "healthy" enteric microbiome? What factors classify a "healthy" enteric microbiome rather than a microbiome associated with malnutrition-related states? Should prophylactic use of rifaximin or metronidazole be utilized in these communities as an alternative preventative therapy for malnutrition? Can specific microbiota-directed foods shape the enteric microbiome and prevent malnutrition states? The answers to these questions can have an impact on millions of vulnerable pediatric lives globally.

While much has been answered in the interaction of the enteric microbiome and obesity and malnutrition, many questions still remain. The intersection of science, medicine, and public health provides opportunities for lasting and significantly impactful interventions to target two of the most pressing issues facing the global community today.

Conclusion

The human body is intricately connected to a world of microorganisms that fluctuate based on our diet, antibiotic use, and nutritional intake. When and how humans begin to develop their microbiome remain controversial. Regardless, if acquisition occurs as a fetus or immediately postnatal, the maternal diet plays a direct role in the development of the neonatal enteric microbiome. Exposure to a maternal high-fat diet leads to persistent changes, specifically decreased commensal species (*Campylobacter*) and increased potentially pathologic bacteria (Bacteroidetes). In addition, a maternal high-fat diet is associated with dysregulation of thyroid hormones and increased anxiety activity in offspring. These findings suggest that it is the high-fat diet and not obesity that leads to alterations in the developing neonatal microbiome. Furthermore, breastfeeding contains more than just milk; HMOs enhance the development of healthy commensal species and even prevent infections from pathogenic organisms. Breast milk itself has its own microbiome that likely derives from enteromammary trafficking from the woman's gastrointestinal system. These findings highlight why breastfeeding has been consistently proven superior than formula feeding for both short- and long-term neonatal outcomes.

Obesity, an ever-pressing problem reaching epidemic proportions throughout the world, is associated with decreased diversity of the gut microbiome and dysbiosis—specifically higher levels of Firmicutes and lower levels of Bacteroidetes. Higher levels of Firmicutes, such as *Lactobacillus*, are also associated with markers of systemic inflammation, suggesting an association between the enteric microbiome and systemic inflammation in obese individuals. Further underscoring how early dysregulation of the developing microbiome may lead to later development of childhood obesity, repeated exposures to broad-spectrum antibiotics has consistently been found to be associated with obesity.

Severe malnutrition expressed as extremely low BMI is associated with its own perturbations of the microbiome. Mice given fecal transplants from infants suffering from kwashiorkor are more likely to lose weight, have higher levels of bacteria associated with intestinal inflammation, and even show inhibition of energy-production pathways preventing successful treatment of the underlying malnutrition. Chronic malnutrition, specifically stunting, is consistently associated with dysbiosis of the enteric microbiome, specifically SIBO. However, further studies are needed to evaluate treatment of SIBO and the amelioration of micro- and macronutrient deficiencies.

Maternal and pediatric nutritional intake plays an integral role in not only growth and maturation, but also in the proper development of a healthy, diverse enteric microbiome. Obesity and malnutrition both have associated dysbiosis with increased inflammatory markers showcasing how intricately connected nutritional status is to the human microbiome and overall health. However, it remains unclear whether directly altering the enteric microbiome during the early, critically important maturational window can prevent the development of obesity or malnutrition later in life. Thus further robust, scientific studies are needed to expand on these findings and to evaluate the impact of direct alterations on the budding microbiome and later development of nutrition-related diseases and disorders.

References

[1] Sender R, Fuchs S, Milo R. Revised estimates for the number of human and bacteria cells in the body. PLoS Biol 2016. Available from: https://doi.org/10.1371/journal.pbio.1002533.

[2] Barker DJP, et al. Weight in infancy and death from ischaemic heart disease. Lancet 1989. Available from: https://doi.org/10.1016/S0140-6736(89)90710-1.

[3] Barker DJP, et al. Fetal nutrition and cardiovascular disease in adult life. Lancet 1993. Available from: https://doi.org/10.1016/0140-6736(93)91224-A.

[4] Barker DJP. The origins of the developmental origins theory. J Intern Med 2007. Available from: https://doi.org/10.1111/j.1365-2796.2007.01809.x.

[5] Barker DJP, Osmond C. Infant mortality, childhood nutrition, and ischaemic heart disease in england and wales. Lancet 1986. Available from: https://doi.org/10.1016/S0140-6736(86)91340-1.

[6] Aagaard-Tillery KM, et al. Developmental origins of disease and determinants of chromatin structure: maternal diet modifies the primate fetal epigenome. J Mol Endocrinol 2008. Available from: https://doi.org/10.1677/JME-08-0025.

[7] Bocock PN, Aagaard-Tillery KM. Animal models of epigenetic inheritance. SemReprod Med 2009. Available from: https://doi.org/10.1055/s-0029-1237425.

[8] Ganu RS, et al. Early origins of adult disease: approaches for investigating the programmable epigenome in humans, nonhuman primates, and rodents. ILAR J 2012. Available from: https://doi.org/10.1093/ilar.53.3-4.306.

[9] Stiemsma LT, Michels KB. The role of the microbiome in the developmental origins of health and disease. Pediatrics 2018. Available from: https://doi.org/10.1542/peds.2017-2437.

[10] Suter M, et al. Epigenomics: maternal high-fat diet exposure in utero disrupts peripheral circadian gene expression in nonhuman primates. FASEB J 2011. Available from: https://doi.org/10.1096/fj.10-172080.

[11] Suter MA, Aagaard-Tillery KM. Environmental influences on epigenetic profiles. SemReprod Med 2009. Available from: https://doi.org/10.1055/s-0029-1237426.

[12] Wang Jun, et al. A metagenome-wide association study of gut microbiota in type 2 diabetes. Nature 2012. Available from: https://doi.org/10.1038/nature11450.

[13] Wu X, et al. Molecular characterisation of the faecal microbiota in patients with type II diabetes. Curr Microbiol 2010. Available from: https://doi.org/10.1007/s00284-010-9582-9.

[14] Larsen N, et al. Gut microbiota in human adults with type 2 diabetes differs from non-diabetic adults. PLoS One 2010. Available from: https://doi.org/10.1371/journal.pone.0009085.

[15] Tilg H, Moschen AR. Microbiota and diabetes: an evolving relationship. Gut 2014. Available from: https://doi.org/10.1136/gutjnl-2014-306928.

[16] Cani PD, et al. Selective increases of bifidobacteria in gut microflora improve high-fat-diet-induced diabetes in mice through a mechanism associated with endotoxaemia. Diabetologia 2007. Available from: https://doi.org/10.1007/s00125-007-0791-0.

[17] Ley RE. Obesity and the human microbiome. Curr Opin Gastroenterol 2010. Available from: https://doi.org/10.1097/MOG.0b013e328333d751.

[18] Schwiertz A, et al. Microbiota and SCFA in lean and overweight healthy subjects. Obesity 2010. Available from: https://doi.org/10.1038/oby.2009.167.

[19] Bäckhed F, et al. Mechanisms underlying the resistance to diet-induced obesity in germ-free mice. Proc Natl Acad Sci U S A 2007. Available from: https://doi.org/10.1073/pnas.0605374104.

[20] Turnbaugh PJ, et al. An obesity-associated gut microbiome with increased capacity for energy harvest. Nature 2006. Available from: https://doi.org/10.1038/nature05414.

[21] Turnbaugh PJ, et al. Diet-induced obesity is linked to marked but reversible alterations in the mouse distal gut microbiome. Cell Host Microbe 2008. Available from: https://doi.org/10.1016/j.chom.2008.02.015.

[22] Turnbaugh PJ, Hamady M, et al. A core gut microbiome in obese and lean twins. Nature 2009. Available from: https://doi.org/10.1038/nature07540.

[23] Koren O, et al. Host remodeling of the gut microbiome and metabolic changes during pregnancy. Cell 2012. Available from: https://doi.org/10.1016/j.cell.2012.07.008.

[24] Markle JGM, et al. Sex differences in the gut microbiome drive hormone-dependent regulation of autoimmunity. Science 2013. Available from: https://doi.org/10.1126/science.1233521.

[25] Valentine G, et al. Relationships between perinatal interventions, maternal-infant microbiomes, and neonatal outcomes. Clin Perinatol 2018;45(2). Available from: https://doi.org/10.1016/j.clp.2018.01.008.

[26] Valentine G, Prince A, Aagaard KM. The neonatal microbiome and metagenomics: what do we know and what is the future? NeoReviews 2019. Available from: https://doi.org/10.1542/neo.20-5-e258.

[27] Chu DM, et al. The development of the human microbiome: why moms matter. Gastroenterology Clin North Am 2019. Available from: https://doi.org/10.1016/j.gtc.2019.04.004.

[28] Afshin A, et al. Health effects of overweight and obesity in 195 countries over 25 years. N Engl J Med 2017. Available from: https://doi.org/10.1056/NEJMoa1614362.

[29] Ahmed T, Hossain M, Sanin KI. Global burden of maternal and child undernutrition and micronutrient deficiencies. Ann Nutr Metab 2012. Available from: https://doi.org/10.1159/000345165.

[30] David LA, et al. Diet rapidly and reproducibly alters the human gut microbiome. Nature 2014. Available from: https://doi.org/10.1038/nature12820.

[31] Ma J, et al. High-fat maternal diet during pregnancy persistently alters the offspring microbiome in a primate model. Nat Commun 2014. Available from: https://doi.org/10.1038/ncomms4889.

[32] Suter MA, Chen A, et al. A maternal high-fat diet modulates fetal SIRT1 histone and protein deacetylase activity in nonhuman primates. FASEB J 2012. Available from: https://doi.org/10.1096/fj.12-212878.

[33] Suter MA, Sangi-Haghpeykar H, et al. Maternal high-fat diet modulates the fetal thyroid axis and thyroid gene expression in a nonhuman primate model. Mol Endocrinol 2012. Available from: https://doi.org/10.1210/me.2012-1214.

[34] Suter MA, et al. Postweaning exposure to a high-fat diet is associated with alterations to The hepatic histone code in Japanese macaques. Pediatric Res 2013. Available from: https://doi.org/10.1038/pr.2013.106.

[35] Suter M, Abramovici A, Aagaard-Tillery K. Genetic and epigenetic influences associated with intrauterine growth restriction due to in utero tobacco exposure. Pediatr Endocrinol Rev 2010.

[36] Bassols J, et al. Gestational diabetes is associated with changes in placental microbiota and microbiome. Pediatr Res 2016. Available from: https://doi.org/10.1038/pr.2016.155.

[37] Gomez-Arango LF, et al. Connections between the gut microbiome and metabolic hormones in early pregnancy in overweight and obese women. Diabetes 2016. Available from: https://doi.org/10.2337/db16-0278.

[38] Chu DM, et al. The early infant gut microbiome varies in association with a maternal high-fat diet. Genome Med 2016. Available from: https://doi.org/10.1186/s13073-016-0330-z.

[39] American Academy of Pediatrics. Breastfeeding and the use of human milk. Policy statement. Pediatrics 2012. Available from: https://doi.org/10.1542/peds.2011-3552.

[40] Ip S, et al. Breastfeeding and maternal and infant health outcomes in developed countries. Evid Rep Technol Assess 2007. Available from: https://doi.org/10.1542/gr.18-2-15.

[41] Duijts L, et al. Prolonged and exclusive breastfeeding reduces the risk of infectious diseases in infancy. Pediatrics 2010. Available from: https://doi.org/10.1542/peds.2008-3256.

[42] Ip S, et al. A summary of the agency for healthcare research and quality's evidence report on breastfeeding in developed countries. Breastfeed Med 2009. Available from: https://doi.org/10.1089/bfm.2009.0050.

[43] Quigley MA, Kelly YJ, Sacker A. Breastfeeding and hospitalization for diarrheal and respiratory infection in the United Kingdom millennium cohort study. Pediatrics 2007. Available from: https://doi.org/10.1542/peds.2006-2256.

[44] Sullivan S, et al. An exclusively human milk-based diet is associated with a lower rate of necrotizing enterocolitis than a diet of human milk and bovine milk-based products. J Pediatr 2010. Available from: https://doi.org/10.1016/j.jpeds.2009.10.040.

[45] Bartick M, Reinhold A. The burden of suboptimal breastfeeding in the United States: a pediatric cost analysis. Pediatrics 2010. Available from: https://doi.org/10.1542/peds.2009-1616.

[46] Chen A, Rogan WJ. Breastfeeding and the risk of postneonatal death in the United States. Pediatrics 2004. Available from: https://doi.org/10.1542/peds.113.5.e435.

[47] Hauck FR, et al. Breastfeeding and reduced risk of sudden infant death syndrome: a meta-analysis. Pediatrics 2011. Available from: https://doi.org/10.1542/peds.2010-3000.

[48] Moon RY, et al. SIDS and other sleep-related infant deaths: expansion of recommendations for a safe infant sleeping environment. Pediatrics 2011. Available from: https://doi.org/10.1542/peds.2011-2285.

[49] Jones G, et al. How many child deaths can we prevent this year? Lancet 2003. Available from: https://doi.org/10.1016/S0140-6736(03)13811-1.

[50] Pannaraj PS, et al. Association between breast milk bacterial communities and establishment and development of the infant gut microbiome. JAMA Pediatr 2017. Available from: https://doi.org/10.1001/jamapediatrics.2017.0378.

[51] Kunz C. Historical aspects of human milk oligosaccharides. Adv Nutr 2012. Available from: https://doi.org/10.3945/an.111.001776.

[52] Schonfeld H. Uber die Beziehungen der einzelnen Bestandteile der Frauenmilch zur Bifidusflora. Jahrb der Kinderh 1926;113:19–60.

[53] Thurl S, et al. Variation of human milk oligosaccharides in relation to milk groups and lactational periods. Br J Nutr 2010. Available from: https://doi.org/10.1017/S0007114510002072.

[54] Hong P, et al. Human milk oligosaccharides reduce HIV-1-gp120 binding to dendritic cell-specific ICAM3-grabbing non-integrin (DC-SIGN). Br J Nutr 2009. Available from: https://doi.org/10.1017/S0007114508025804.

[55] Morozov V, et al. Human milk oligosaccharides as promising antivirals. Mol Nutr Food Res 2018. Available from: https://doi.org/10.1002/mnfr.201700679.

[56] Hunt KM, et al. Characterization of the diversity and temporal stability of bacterial communities in human milk. PLoS One 2011. Available from: https://doi.org/10.1371/journal.pone.0021313.

[57] Khodayar-Pardo P, et al. Impact of lactation stage, gestational age and mode of delivery on breast milk microbiota. J Perinatol 2014. Available from: https://doi.org/10.1038/jp.2014.47.

[58] LaTuga MS, Stuebe A, Seed PC. A review of the source and function of microbiota in breast milk. Semin Reprod Med 2014. Available from: https://doi.org/10.1055/s-0033-1361824.

[59] Jiménez E, et al. Oral administration of *Lactobacillus* strains isolated from breast milk as an alternative for the treatment of infectious mastitis during lactation. Appl Environ Microbiol 2008. Available from: https://doi.org/10.1128/AEM.02599-07.

[60] Hair AB, et al. Beyond necrotizing enterocolitis prevention: improving outcomes with an exclusive human milk-based diet. Breastfeed Med 2016. Available from: https://doi.org/10.1089/bfm.2015.0134.

[61] Gopalakrishna KP, et al. Maternal IgA protects against the development of necrotizing enterocolitis in preterm infants. Nat Med 2019. Available from: https://doi.org/10.1038/s41591-019-0480-9.

[62] Skinner AC, Perrin EM, Skelton JA. Prevalence of obesity and severe obesity in US children, 1999–2014. Obesity 2016. Available from: https://doi.org/10.1002/oby.21497.

[63] Biro FM, Wien M. Childhood obesity and adult morbidities. Am J Clin Nutr 2010. Available from: https://doi.org/10.3945/ajcn.2010.28701B.

[64] Armstrong SC, et al. Pediatric metabolic and bariatric surgery: evidence, barriers, and best practices. Pediatrics 2019. Available from: https://doi.org/10.1542/peds.2019-3223.

[65] Ley RE, et al. Obesity alters gut microbial ecology. Proc Natl Acad Sci U S A 2005. Available from: https://doi.org/10.1073/pnas.0504978102.

[66] Turnbaugh PJ, Ridaura VK, et al. The effect of diet on the human gut microbiome: a metagenomic analysis in humanized gnotobiotic mice. Sci Transl Med 2009. Available from: https://doi.org/10.1126/scitranslmed.3000322.

[67] Waldram A, et al. Top-down systems biology modeling of host metabotype-microbiome associations in obese rodents. J Proteome Res 2009. Available from: https://doi.org/10.1021/pr8009885.

[68] Murphy EF, et al. Composition and energy harvesting capacity of the gut microbiota: relationship to diet, obesity and time in mouse models. Gut 2010. Available from: https://doi.org/10.1136/gut.2010.215665.

[69] Ley RE, et al. Microbial ecology: human gut microbes associated with obesity. Nature 2006. Available from: https://doi.org/10.1038/4441022a.

[70] Scheepers LEJM, et al. The intestinal microbiota composition and weight development in children: The KOALA Birth Cohort Study. Int J Obes 2015. Available from: https://doi.org/10.1038/ijo.2014.178.

[71] Vael C, et al. Intestinal microflora and body mass index during the first three years of life: an observational study. Gut Pathog 2011. Available from: https://doi.org/10.1186/1757-4749-3-8.

[72] Bervoets L, et al. Differences in gut microbiota composition between obese and lean children: a cross-sectional study. Gut Pathog 2013. Available from: https://doi.org/10.1186/1757-4749-5-10.

[73] Million M, et al. Obesity-associated gut microbiota is enriched in *Lactobacillus reuteri* and depleted in *Bifidobacterium* animalis and *Methanobrevibacter smithii*. Int J Obes 2012. Available from: https://doi.org/10.1038/ijo.2011.153.

[74] Ismail NA, et al. Frequency of Firmicutes and Bacteroidetes in gut microbiota in obese and normal weight Egyptian children and adults. Arch Med Sci 2011. Available from: https://doi.org/10.5114/aoms.2011.23418.

[75] Lu L, et al. Transcriptional modulation of intestinal innate defense/inflammation genes by preterm infant microbiota in a humanized gnotobiotic mouse model. PLoS One 2015. Available from: https://doi.org/10.1371/journal.pone.0124504.

[76] Bailey LC, et al. Association of antibiotics in infancy with early childhood obesity. JAMA Pediatr 2014. Available from: https://doi.org/10.1001/jamapediatrics.2014.1539.

[77] Neu J. The microbiome and its impact on disease in the preterm patient. Curr Pediatr Rep 2013. Available from: https://doi.org/10.1007/s40124-013-0031-7.

[78] Jernberg C, et al. Long-term impacts of antibiotic exposure on the human intestinal microbiota. Microbiology 2010. Available from: https://doi.org/10.1099/mic.0.040618-0.

[79] Azad MB, et al. Pediatric original article infant antibiotic exposure and the development of childhood overweight and central adiposity. Int J Obes 2014. Available from: https://doi.org/10.1038/ijo.2014.119.

[80] Ajslev TA, et al. Childhood overweight after establishment of the gut microbiota: the role of delivery mode, pre-pregnancy weight and early administration of antibiotics. Int J Obes 2011. Available from: https://doi.org/10.1038/ijo.2011.27.

[81] Black RE, et al. Maternal and child undernutrition and overweight in low-income and middle-income countries. Lancet 2013. Available from: https://doi.org/10.1016/S0140-6736(13)60937-X.

[82] Bhutta ZA, et al. What works? Interventions for maternal and child undernutrition and survival. Lancet 2008. Available from: https://doi.org/10.1016/S0140-6736(07)61693-6.

[83] Caulfield LE, et al. Undernutrition as an underlying cause of child deaths associated with diarrhea, pneumonia, malaria, and measles. Am J Clin Nutr 2004. Available from: https://doi.org/10.1093/ajcn/80.1.193.

[84] Kau AL, et al. Human nutrition, the gut microbiome and the immune system. Nature 2011. Available from: https://doi.org/10.1038/nature10213.
[85] De Filippo C, et al. Impact of diet in shaping gut microbiota revealed by a comparative study in children from Europe and rural Africa. Proc Natl Acad Sci U S A 2010. Available from: https://doi.org/10.1073/pnas.1005963107.
[86] Yatsunenko T, et al. Human gut microbiome viewed across age and geography. Nature 2012. Available from: https://doi.org/10.1038/nature11053.
[87] Raman AS, et al. A sparse covarying unit that describes healthy and impaired human gut microbiota development. Science 2019. Available from: https://doi.org/10.1126/science.aau4735.
[88] Subramanian S, et al. Persistent gut microbiota immaturity in malnourished Bangladeshi children. Nature 2014. Available from: https://doi.org/10.1038/nature13421.
[89] Gehrig JL, et al. Effects of microbiota-directed foods in gnotobiotic animals and undernourished children. Science 2019. Available from: https://doi.org/10.1126/science.aau4732.
[90] Blanton LV, et al. Gut bacteria that prevent growth impairments transmitted by microbiota from malnourished children. Science 2016. Available from: https://doi.org/10.1126/science.aad3311.
[91] Smith MI, et al. Gut microbiomes of Malawian twin pairs discordant for kwashiorkor. Science 2013. Available from: https://doi.org/10.1126/science.1229000.
[92] Galloway R. Global nutrition outcomes at ages 5 to 19. 3rd ed. Dis control priorities, vol. 8. Child and adolescent health and development; 2017. Available from: https://do.org/10.1596/978-1-4648-0423-6_ch3.
[93] Vonaesch P, et al. Stunted childhood growth is associated with decompartmentalization of the gastrointestinal tract and overgrowth of oropharyngeal taxa. Proc Natl Acad Sci U S A 2018. Available from: https://doi.org/10.1073/pnas.1806573115.
[94] Donowitz JR, et al. Small intestine bacterial overgrowth and environmental enteropathy in Bangladeshi children. mBio 2016. Available from: https://doi.org/10.1128/mBio.02102-15.
[95] Bures J, et al. Small intestinal bacterial overgrowth syndrome. World J Gastroenterol 2010. Available from: https://doi.org/10.3748/wjg.v16.i24.2978.
[96] Galloway D, et al. Serum unconjugated bile acids and small bowel bacterial overgrowth in pediatric intestinal failure: a pilot study. J Parenter Enter Nutr 2019. Available from: https://doi.org/10.1002/jpen.1316.
[97] Campbell DI, Elia M, Lunn PG. Growth faltering in rural Gambian infants is associated with impaired small intestinal barrier function, leading to endotoxemia and systemic inflammation. J Nutr 2003. Available from: https://doi.org/10.1093/jn/133.5.1332.
[98] Canani RB, et al. Potential beneficial effects of butyrate in intestinal and extraintestinal diseases. World J Gastroenterol 2011. Available from: https://doi.org/10.3748/wjg.v17.i12.1519.
[99] Cole CR, et al. The rate of bloodstream infection is high in infants with short bowel syndrome: relationship with small bowel bacterial overgrowth, enteral feeding, and inflammatory and immune responses. J Pediatr 2010. Available from: https://doi.org/10.1016/j.jpeds.2009.12.008.
[100] De Boissieu D, et al. Small-bowel bacterial overgrowth in children with chronic diarrhea, abdominal pain, or both. J Pediatr 1996. Available from: https://doi.org/10.1016/S0022-3476(96)70390-6.

[101] Collins BS, Lin HC. Double-blind, placebo-controlled antibiotic treatment study of small intestinal bacterial overgrowth in children with chronic abdominal pain. J Pediatr Gastroenterol Nutr 2011. Available from: https://doi.org/10.1097/MPG.0b013e3181effa3b.

[102] Korterink JJ, et al. Glucose hydrogen breath test for small intestinal bacterial overgrowth in children with abdominal pain-related functional gastrointestinal disorders. J Pediatr Gastroenterol Nutr 2015. Available from: https://doi.org/10.1097/MPG.0000000000000634.

[103] Siniewicz-Luzenczyk K, et al. Small intestinal bacterial overgrowth syndrome in children. Prz Gastroenterol 2015. Available from: https://doi.org/10.5114/pg.2014.47494.

[104] Shah SC, et al. Meta-analysis: antibiotic therapy for small intestinal bacterial overgrowth. Aliment Pharmacol Ther 2013. Available from: https://doi.org/10.1111/apt.12479.

[105] Gatta L, et al. Systematic review with meta-analysis: rifaximin is effective and safe for the treatment of small intestine bacterial overgrowth. Aliment Pharmacol Ther 2017. Available from: https://doi.org/10.1111/apt.13928.

[106] Tahan S, et al. Effectiveness of trimethoprim-sulfamethoxazole and metronidazole in the treatment of small intestinal bacterial overgrowth in children living in a slum. J Pediatr Gastroenterol Nutr 2013. Available from: https://doi.org/10.1097/MPG.0b013e3182952e93.

[107] Quigley EMM, Quera R. Small intestinal bacterial overgrowth: roles of antibiotics, prebiotics, and probiotics. Gastroenterology 2006. Available from: https://doi.org/10.1053/j.gastro.2005.11.046.

Chapter 9

Vitamin D: its impact on disease and the microbiome

Erika C. Claud[1] and Anita Rao[2]
[1]*Departments of Pediatrics and Medicine, Sections of Neonatology and Gastroenterology, The University of Chicago, Chicago, IL, United States,* [2]*Section of Gastroenterology, Hepatology, and Nutrition, Department of Pediatrics, The University of Chicago, Chicago, IL, United States*

Introduction to vitamin D and vitamin D deficiency

Vitamin D is a key nutrient that is involved in calcium and phosphate absorption in the intestine, calcium reabsorption in the kidney, and calcium mobilization in bone. The human body is capable of both synthesizing vitamin D de novo and absorbing it from dietary sources. The majority of vitamin D in the body is synthesized in a multistep process beginning with the skin's exposure to UV-B radiation, resulting in synthesis of vitamin D from 7-dehydrocholesterol. Vitamin D is subsequently converted to the active hormone via the action of 25-hydroxylase in the liver and 1α-hydroxylase in the kidney, forming 1,25$(OH)_2$ vitamin D, also known as "Calcitriol" (Fig. 9.1) [1].

A range of factors, including season and latitude, affect an individual's vitamin D levels. Vitamin D can be absorbed from dietary sources in the small intestine, but few foods naturally contain the nutrient. Fish, egg yolks, cheese, and mushrooms are the only dietary sources that contain natural vitamin D, with fish having the highest concentration of vitamin D. Other significant food sources are fortified with the vitamin, such as cow's milk, orange juice, and margarine [2–4].

Vitamin D is imperative in bone health, and severe vitamin D deficiency results in skeletal deformities. The term "rickets" describes impaired bone mineralization in the growth plates of children, and the term "osteomalacia" describes impaired bone mineralization in the bone matrix of adults [5]. The recommended serum concentration of vitamin D is controversial and not well supported by clinical trials, especially in the pediatric population. Vitamin D status is generally measured by 25(OH) vitamin D levels because of its long half-life of 2–3 weeks and more robust circulating concentration [6]. In 2017, international experts in pediatric endocrinology, nutrition, and

The Developing Microbiome. DOI: https://doi.org/10.1016/B978-0-12-820602-7.00009-X

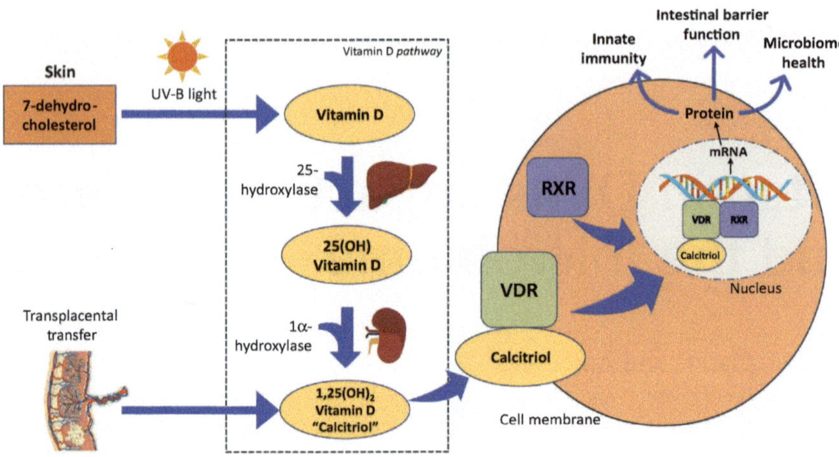

FIGURE 9.1 Pathway depicting vitamin D synthesis and action in target cells. Vitamin D is synthesized via UV-B light irradiation on the skin, which converts 7-dehydrocholesterol to vitamin D. Vitamin D then undergoes serial hydroxylations in the liver and kidney to form active 1,25(OH)$_2$ vitamin D (also known as calcitriol). In utero, calcitriol is transferred to the fetus via the placenta, and this process accounts for the entirety of a newborn infant's vitamin D stores. Calcitriol then binds to the vitamin D receptor (VDR) on the cell membrane. Upon binding to its ligand, VDR complexes with the retinoid X receptor (RXR) and translocates to the nucleus of the cell. Here it binds to vitamin D responsive elements (VDRE) of target genes and acts as a transcription factor for the synthesis of proteins. Products of vitamin D signaling have a wide range of roles influencing innate immunity, intestinal barrier function, and microbiome health.

public health representing 10 pediatric societies released the *Global Consensus Recommendations on the Prevention and Management of Nutritional Rickets*, which defined vitamin D insufficiency as 12–20 ng/mL and deficiency as a concentration less than 12 ng/mL in children [7].

For newborns, vitamin D status is dependent on maternal vitamin D concentration and its ability to be transported across the placenta [8]. Breast milk has comparatively lower amounts of vitamin D compared to formula, and thus exclusively breastfed infants receive below the recommended intake of vitamin D without supplementation. In fact, a recent systematic review found that neonatal vitamin D deficiency is a widespread issue in many countries [9]. To combat deficiency, the American Academy of Pediatrics (AAP) recommends supplementing all exclusively breastfed infants with 400 IU of vitamin D per day [10].

Vitamin D is one of the most prescribed medications in the neonatal intensive care unit [11]. Preterm infants are more likely to be vitamin D deficient due to the shorter duration of placental transfer of vitamin D from the mother, prolonged hospitalizations leading to poor UV-B exposure, and decreased fat reserve leading to less vitamin storage. To address this deficiency, the AAP published an expert opinion report in 2013 recommending

200–400 IU per day of vitamin D supplementation in very low-birthweight infants (weighing less than 1500 g) and 400 IU per day of supplementation in infants weighing more than 1500 g [10]. Additional research has shown that despite breast milk being low in vitamin D, maternal supplementation with 2000–4000 IU of vitamin D per day was equivalent to oral supplementation of exclusively breastfed newborns with 400 IU per day [12–14]. The purpose of this review is to explore the roles of vitamin D beyond skeletal health (particularly in infants), as well as its potential immunomodulatory properties, with a specific focus on its interaction with the microbiome.

Vitamin D receptor

In order to better understand the downstream effects of vitamin D, one must understand the actions of the vitamin D receptor (VDR). VDR is a member of the nuclear receptor superfamily and is highly expressed in the intestines, kidneys, parathyroid glands, and bones [15]. Most of the actions of the vitamin D axis involve gene regulation following the binding of VDR to its ligand ($1,25(OH)_2D$) which then binds to the retinoid-X receptor (RXR), forming a VDR–vitamin D–RXR complex. This complex then binds vitamin D-responsive elements located in the promoter regions of target genes, thus functioning as a transcription factor (Fig. 9.1) [16]. More than 50 target genes of VDR have been identified that have a broad range of actions including cellular proliferation and have an important role in the innate immune system as well as the intestinal microbiome, to be discussed in detail later in the chapter.

Vitamin D and infancy

Vitamin D has been shown to be immensely beneficial in promoting growth and development in infants and young children. It is well-known that vitamin D prevents rickets, and it has also been shown to aid in neurodevelopment by promoting neurite growth and by inhibiting neuronal apoptosis [17]. Conversely, vitamin D deficiency in infants/young children is implicated in atopic disease, including asthma, allergies, and eczema [18–20]. Interestingly, although the exact pathogenesis of atopic disease is not known, many studies have postulated an association between atopy and an altered microbiome, including decreased microbial diversity [21–25].

Cow's milk protein allergy (CMPA) is the most prevalent type of food allergy in childhood, with an incidence of 2%–3% in the first year of life. CMPA describes eosinophil-dominant intestinal inflammation with or without increased blood eosinophilia in the presence of cow's milk proteins, including whey and casein. In one study, low serum vitamin D levels were associated with persistence of blood eosinophilia, as well as symptoms in infants with CMPA, which shows a potential link between CMPA and vitamin D deficiency [20].

The Vitamin D Antenatal Asthma Reduction Trial (VDAART) was developed to specifically investigate if prenatal vitamin D supplementation is able to prevent atopic disease in young children. VDAART is a randomized controlled trial of vitamin D supplementation in ethnically diverse pregnant women. As part of this study, cord blood vitamin D levels were obtained from enrolled infants, and the infants were followed prospectively at 3 years of age to determine whether they had developed atopic disease. Based on the results of this study, maternal supplementation with 4400 IU of vitamin D per day resulted in higher maternal vitamin D levels as well as higher cord blood vitamin D levels compared to maternal supplementation with 400 IU per day. However, there was no significant decrease in the incidence of atopy between the two treatment groups [26]. This shows the need for more research on the potential role of vitamin D and atopic disease in infants/young children. Vitamin D deficiency in mothers/neonates has also been seen as a risk factor for neonatal sepsis [27]. In one single-center randomized controlled trial of vitamin D supplementation for premature infants with late-onset sepsis, the majority of infants enrolled were found to be vitamin D deficient. The infants were randomly assigned to treatment groups receiving either 400 or 800 IU of vitamin D per day. For both treatment groups, serum levels of pro-inflammatory cytokines including TNF-α and IL-6 decreased, with no significant difference between the groups [28]. This highlights a potential antiinflammatory role of vitamin D. Finally, maternal vitamin D deficiency has been implicated in preterm birth and prematurity [29]. These findings underscore the importance of vitamin D in growth and development in the perinatal and postnatal periods.

Vitamin D and inflammatory bowel diseases

Recent studies have shown a potential link between intestinal inflammation and vitamin D signaling. VDR has been found to be intestinally protective in numerous studies, using both intestinal cell lines and mouse models [30–32]. VDR protein expression in the small intestine has been shown to be a positive regulator of microvilli formation [33]. VDR expression in the colon has been found to protect against colonic inflammation by decreasing epithelial cell apoptosis and by strengthening tight junction proteins, thus decreasing intestinal permeability [31,34,35]. The role of VDR in innate immunity and antimicrobial defense has been well-described, with intestinal epithelial VDR knockout mice having diminished Paneth cell populations and more susceptibility to dextran sulfate sodium (DSS)-induced colitis [36]. Paneth cells are intestinal epithelial cells located in the small intestine that are responsible for secreting antimicrobial peptides (AMPs) such as cathelicidin and β-defensin, thus playing a key role in the innate immune system of the gut [15,36]. Vitamin D signaling has also been shown to inhibit the production of pro-inflammatory cytokines produced by T-helper cells and by the NF-κB signaling pathway including IL-2, IFN-γ, and TNF-α [28]. This underscores the importance of vitamin D in innate immunity.

A growing body of literature has associated a disrupted vitamin D/VDR pathway with both increased risk and increased severity of inflammatory bowel disease (IBD) [37–39]. IBDs, which includes Crohn's disease (CD) and ulcerative colitis (UC), are chronic, recurrent diseases of the gastrointestinal tract. Although the exact etiology of IBD is unknown, disease pathogenesis is likely due to a combination of genetic predisposition, intestinal dysbiosis, and immune system dysregulation [40].

A meta-analysis summarized four polymorphisms (*TaqI*, *BsmI*, *FokI*, and *ApaI*) in the VDR gene that are associated with increased susceptibility to CD and UC [41]. Furthermore, vitamin D deficiency has been associated with higher disease activity scores [42,43] and objective markers of disease severity such as erythrocyte sedimentation rate (ESR) levels, C-reactive peptide (CRP) levels, fecal calprotectin, and the presence of endoscopic inflammation [44,45].

The potentially protective role of the vitamin D/VDR pathway thus raises questions about whether vitamin D supplementation has therapeutic benefits in intestinal disease. Vitamin D supplementation was found to be protective against lipopolysaccharide (LPS)-induced intestinal inflammation in Caco-2 cells [31]. Similarly, mice fed a vitamin D-deficient diet were found to be more susceptible to *Citrobacter rodentium* colitis than mice fed a diet rich in vitamin D [46]. Related to this, treatment with a VDR ligand lithocholic acid rescued TNF-α induced intestinal injury in Caco-2 cells [30], showing that VDR activation with either vitamin D or a vitamin D analog protects against intestinal injury. Incubating peripheral and intestinal T cells from IBD patients with vitamin D resulted in decreased frequencies of pro-inflammatory CD4+ and CD8+ T cells. Furthermore, normalization of serum vitamin D concentrations in previously deficient CD patients resulted in significantly reduced levels of peripheral CD4+ T cells producing IFN-γ, IL-17, and IL-22. This suggests a potential antiinflammatory role of vitamin D supplementation [47].

Vitamin D deficiency is a common issue in individuals with IBD; however, studies are conflicted on the role of vitamin D supplementation as a therapeutic agent in IBD. In a randomized double-blind controlled trial of 108 patients with Crohn's disease in remission, daily oral supplementation with 1200 IU vitamin D reduced the risk of relapse from 29% to 13%; however, this result was not statistically significant [43]. In another randomized double-blind controlled trial, a single IM injection of 7.5 mg vitamin D was effective in raising serum vitamin D levels and in decreasing the levels of inflammatory cytokines such as TNF-α and IFN-γ after 90 days. This study reached statistical significance [48]. Conversely, other studies showed that in subjects with IBD, vitamin D supplementation improved serum vitamin D levels but did not improve objective markers of disease activity such as fecal calprotectin, levels of inflammatory markers, or rate of relapse compared to placebo [49,50]. In a meta-analysis, vitamin D supplementation was seen to improve vitamin D levels and reduce the rate of relapse in IBD subjects compared to placebo. However, vitamin D did not significantly improve objective markers of disease activity such as ESR and CRP levels in these subjects [51]. More research, including larger randomized controlled trials, is therefore needed on the potential role of vitamin D as a therapeutic adjuvant in the treatment of IBD.

One specific note must be made in regard to the role of vitamin D deficiency in the pathogenesis of necrotizing enterocolitis (NEC), which is an inflammatory bowel disease of premature infants characterized by inflammation and bacterial invasion of the bowel wall. The clinical features of NEC include abdominal distension, feeding intolerance, bloody stools, and intestinal perforation. While the exact pathogenesis of NEC is unknown, it is likely due to a combination of factors including genetics, intestinal dysbiosis, and high immunoreactivity [52,53].

Toll-like receptor 4 (TLR4) is an inflammatory cytokine that plays a key role in the development of NEC. Preterm infants with NEC were found to have significantly lower serum vitamin D levels than preterm infants without NEC. In an animal model using neonatal Wistar rats, NEC was induced through formula feeding as well as cold/asphyxia stress. VDR expression was decreased, and TLR4 expression was increased in this model. Interestingly, treatment with oral vitamin D significantly suppressed TLR4 expression [54], showing that vitamin D may be a negative transcriptional regulator of TLR4. These results implicate vitamin D as a protective factor against NEC, potentially due to its immunomodulatory properties as well its potential effects on the microbiome.

Vitamin D, vitamin D receptor, and the microbiome

Vitamin D is beneficial in the newborn period across many conditions, including for its roles in neurodevelopment, sepsis, atopic disease, and NEC, all of which have interestingly been closely tied to the health of an individual's microbiome [25,52,55,56]. Beyond the newborn period, vitamin D is being extensively studied for its role in IBD, which is also a disease that is linked to intestinal dysbiosis [40]. It has thus been hypothesized that vitamin D signaling is associated with the establishment and maintenance of a healthy gut microbiome. Absence of vitamin D signaling results in dysbiosis, with VDR knockout mice having enrichment of the genera *Clostridium* and *Bacteroides* as well as depletion of *Lactobacillus* in their stool compared to wild-type mice. This correlates vitamin D signaling and VDR activity with the presence of bacteria that produce lactic acid, which lower intestinal pH thereby promoting proper microbial balance [57]. In another study, VDR knockout mice were seen to have higher numbers of the phylum *Proteobacteria* and lower numbers of the family *Lachnospiraceae* from the *Firmicutes* phylum, which is a pattern described in the colon of IBD patients compared to healthy controls [58].

Supporting the idea that an intact vitamin D/VDR pathway is necessary for a healthy gut microbiome, the probiotic VSL #3 was shown to protect wild-type mice from *Salmonella*-induced colitis; however, this protective benefit was not seen in VDR knockout mice. VSL #3 was furthermore shown to increase VDR expression in an IL-10 knockout colitis mouse model [59]. These results suggest that an intact VDR pathway is necessary for probiotic benefit in preventing colitis. This growing body of research additionally suggests a possibly therapeutic role of vitamin D in establishing and/or maintaining a healthy gut microbiome.

As an ancillary study to the previously described VDAART trial, microbiome analysis using 16S rRNA sequencing was performed on 309 stool samples from enrolled infants aged 3 to 6 months. This study showed an association between cord blood vitamin D levels and infant gut microbial composition and abundance. On average, the infant gut microbiome was dominated by gram-negative anaerobes (*Bacteroides*, *Veillonella*), gram-positive anaerobes (*Bifidobacterium*), and facultative microorganisms (*Escherichia/Shigella*). In infants with higher cord blood Vitamin D levels, there was an increased abundance of Unclassified *Clostridiales*, the family *Lachnospiraceae*, and the genus *Lachnobacterium*. These infants also had deceased abundance of the genus *Lactococcus* compared to infants with lower cord blood Vitamin D levels [19].

In the KOALA Birth Cohort Study, serum vitamin D levels were obtained from enrolled mothers at 36 weeks of gestation, and infant fecal samples were obtained at 1 month of age for microbial analysis using 16S rRNA sequencing. The objective of this study was to assess if vitamin D supplementation (maternal and infant) influenced infant gut microbial composition. According to this study, increased maternal vitamin D supplementation and higher maternal vitamin D levels led to increased prevalence of *Bacteroides fragilis* and decreased prevalence of *Bifidobacterium* and *Clostridium difficile* in the infant microbiome. There was no influence of infant vitamin D supplementation during the first month of life on the prevalence of any bacterial species [60].

Looking beyond the neonatal period, vitamin D supplementation was shown to significantly influence microbial communities in adult subjects with CD compared to age-matched healthy controls. Furthermore, the changes seen in the microbiome reflected an increased population of healthy gut bacteria that are associated with decreased IBD activity. *Bacteroidetes* and *Clostridia* were among the most abundant phyla/classes in both CD subjects and healthy controls. In subjects with CD, administration of vitamin D led to increased abundance of *Alistipes* (of the phylum *Bacteroidetes*) and *Parabacteroides* (of the phylum *Bacteroides*) [61]. Both species appear to be of importance in the pathogenesis of IBD, with studies showing a protective role of each in murine models of DSS-induced colitis [62,63].

In another randomized controlled trial of adult subjects, overweight or obese adults with vitamin D deficiency were given vitamin D supplementation versus placebo to assess the effects of supplementation on the fecal microbiome. After 16 weeks of supplementation, the fecal microbiome was analyzed using 16S rRNA sequencing. While there was no significant difference seen in microbial α-diversity between vitamin D and placebo groups during the study period, there was a significant difference seen in microbial composition at the genus level with vitamin D supplementation, with the vitamin D treatment group having a significantly higher abundance of genus *Lachnospira* and lower abundance of genus *Blautia* compared to the placebo group. This shows that vitamin D supplementation may have distinct effects on the microbiome [64].

One proposed mechanism for vitamin D affecting the microbiome takes into account its protective functions, with vitamin D upregulating innate immunity (including the production of AMPs) and maintaining intestinal barrier function. This in turn decreases intestinal inflammation, thus preventing an abundance of pathogenic gut bacteria from establishing a dysbiotic state [65]. A growing body of evidence also suggests that gut microbiota play an essential role in host health by protecting intestinal epithelial cells from injury and promoting innate immunity [66,67]. Therefore, it is possible that vitamin D exerts its immunomodulatory effects through its influence on the microbiome. Current literature, while limited, thus points toward an interplay between vitamin D, the microbiome, and the innate immune system in impacting numerous disease states across an individual's lifespan, including atopic disease, NEC, and IBD.

Conclusion

While research on the effects of the vitamin D/VDR pathway is ongoing, vitamin D has been shown to be impactful in human health across the life course. Studies have shown an immunomodulatory effect of this pathway and a beneficial influence on the balance and diversity of intestinal microorganisms. Vitamin D has shown to be protective against intestinal inflammation, atopic disease, and sepsis in the neonatal period, as well as preterm birth/prematurity. This underscores the importance of prenatal supplementation with vitamin D, as a newborn infant's vitamin D status is dependent maternal vitamin D status.

Beyond the neonatal period, vitamin D is imperative in intestinal health and promotes innate immunity. Numerous studies have revealed the effects of this pathway on IBD. Given that many of these diseases are attributed in part to intestinal dysbiosis, it is possible that the effects of vitamin D on disease activity are modulated through its influence on the microbiome. However, more research is needed on how specifically the vitamin D/VDR pathway impacts and/or is impacted by the microbiome, the specific roles of this pathway in human intestinal disease such as IBD, and the possible therapeutic benefits of vitamin D as well as vitamin D analogs. Given the likely imperative role of vitamin D/VDR in eubiosis, future research will likely be focusing on VDR potentiation and ways to optimize intestinal VDR expression.

References

[1] Nielsen OH, Rejnmark L, Moss AC. Role of vitamin D in the natural history of Inflammatory bowel disease. J Crohns Colitis 2018;12(6):742–52.

[2] Holick MF. Sunlight and vitamin D for bone health and prevention of autoimmune diseases, cancers, and cardiovascular disease. Am J Clin Nutr 2004;80(Suppl):1678S–88S.

[3] Lamberg-Allardt C. Vitamin D in foods and as supplements. Prog Biophys Mol Biol 2006;92:33–8.

[4] Holden JM, Lemar LE. Assessing vitamin D contents in foods and supplements: challenges and needs. Am J Clin Nutr 2008;88(Suppl):551S–3S.

[5] Sahay M, Sahay R. Rickets-vitamin D deficiency and dependency. Indian J Endocrinol Metab 2012;16(2):164–76.

[6] Lee JY, So TY, Thackray J. A review on vitamin D deficiency treatment in pediatric patients. J Pediatr Pharmacol Ther 2013;18(4):277–91.

[7] Munns CF, Shaw N, Kiely M, et al. Global consensus recommendations on prevention and management of nutritional rickets. J Clin Endocrinol Metab 2016;101(2):394–415.

[8] Josefson JL, Feinglass J, Rademaker AW, et al. Maternal obesity and vitamin D sufficiency are associated with cord blood vitamin D insufficiency. J Clin Endocrinol Metab 2013;98(1):114–19.

[9] Dawodu A, Agarwal M, Hossain M, Kochiyil J, Zayed R. Hypovitaminosis D and vitamin D deficiency in exclusively breast-feeding infants and their mothers in summer: a justification for vitamin D supplementation of breast-feeding infants. J Pediatr 2003;142 (2):169–73.

[10] Abrams SA. Committee on nutrition calcium and vitamin D requirements of enterally fed preterm infants. Pediatrics 2013;131(5):e1676–83.

[11] Mesek I, Nellis G, Lass J, et al. Medicines prescription patterns in European neonatal units. Int J Clin Pharm 2019;41(6):1578–91.

[12] Dawodu A, Salameh KM, Al-Janahi NS, Bener A, Elkum N. The effect of high-dose postpartum maternal vitamin D supplementation alone compared with maternal plus infant vitamin D supplementation in breastfeeding infants in a high-risk population. A randomized controlled trial. Nutrients 2019;11(7).

[13] Hollis BW, Wagner CL, Howard CR, et al. Maternal versus infant vitamin D supplementation during lactation: a randomized controlled trial. Pediatrics 2015;136(4):625–34.

[14] Gallo S, McDermid JM, Al-Nimr RI, et al. Vitamin D supplementation during pregnancy: an evidence analysis center systematic review and meta-analysis. J Acad Nutr Diet 2019;.

[15] Wang Y, Zhu J, DeLuca HF. Where is the vitamin D receptor? Arch Biochem Biophys 2012;523(1):123–33.

[16] Plum LA, DeLuca HF. Vitamin D, disease and therapeutic opportunities. Nat Rev Drug Discov 2010;9(12):941–55.

[17] Cerdo T, Dieguez E, Campoy C. Infant growth, neurodevelopment and gut microbiota during infancy: which nutrients are crucial? Curr Opin Clin Nutr Metab Care 2019;22 (6):434–41.

[18] du Toit G, Elizur A, Nadeau KC. Cow's milk and vitamin D supplementation in infants-timing is everything. JAMA Pediatr 2019;.

[19] Sordillo JE, Zhou Y, McGeachie MJ, et al. Factors influencing the infant gut microbiome at age 3-6 months: findings from the ethnically diverse Vitamin D Antenatal Asthma Reduction Trial (VDAART). J Allergy Clin Immunol 2017;139(2):482–91.

[20] Li J, Mei X, Cai X, et al. Association of blood eosinophilia and vitamin D insufficiency in young infants with cow milk allergy. Asia Pac J Clin Nutr 2019;28(3):550–7.

[21] Azad MB, Konya T, Guttman DS, et al. Infant gut microbiota and food sensitization: associations in the first year of life. Clin Exp Allergy 2015;45(3):632–43.

[22] Bunyavanich S, Shen N, Grishin A, et al. Early-life gut microbiome composition and milk allergy resolution. J Allergy Clin Immunol 2016;138(4):1122–30.

[23] Ling Z, Li Z, Liu X, et al. Altered fecal microbiota composition associated with food allergy in infants. Appl Env Microbiol 2014;80(8):2546–54.

[24] Thompson-Chagoyan OC, Fallani M, Maldonado J, et al. Faecal microbiota and short-chain fatty acid levels in faeces from infants with cow's milk protein allergy. Int Arch Allergy Immunol 2011;156(3):325–32.

[25] Dong P, Feng JJ, Yan DY, Lyu YJ, Xu X. Early-life gut microbiome and cow's milk allergy - a prospective case - control 6-month follow-up study. Saudi J Biol Sci 2018;25 (5):875–80.

[26] Litonjua AA, Lange NE, Carey VJ, et al. The Vitamin D Antenatal Asthma Reduction Trial (VDAART): rationale, design, and methods of a randomized, controlled trial of vitamin D supplementation in pregnancy for the primary prevention of asthma and allergies in children. Contemp Clin Trials 2014;38(1):37–50.

[27] Tayel SI, Soliman SE, Elsayed HM. Vitamin D deficiency and vitamin D receptor variants in mothers and their neonates are risk factors for neonatal sepsis. Steroids 2018;134:37–42.

[28] Abdel-Hady H, Yahia S, Megahed A, et al. Mediators in preterm infants with late-onset sepsis. J Pediatric Gastroenterology Nutr 2019;68(4):578–84.

[29] Barchitta M, Maugeri A, La Rosa MC, et al. Single nucleotide polymorphisms in vitamin D receptor gene affect birth weight and the risk of preterm birth: results from the "Mamma & Bambino" cohort and a meta-analysis. Nutrients 2018;10(9).

[30] Yao B, He J, Yin X, Shi Y, Wan J, Tian Z. The protective effect of lithocholic acid on the intestinal epithelial barrier is mediated by the vitamin D receptor via a SIRT1/Nrf2 and NF-kappaB dependent mechanism in Caco-2 cells. Toxicol Lett 2019;316:109–18.

[31] Lee C, Lau E, Chusilp S, et al. Protective effects of vitamin D against injury in intestinal epithelium. Pediatr Surg Int 2019;.

[32] Liu W, Chen Y, Golan MA, et al. Intestinal epithelial vitamin D receptor signaling inhibits experimental colitis. J Clin Invest 2013;123(9):3983–96.

[33] Barbáchano A, Fernández-Barral A, Ferrer-Mayorga G, Costales-Carrera A, Larriba MJ, Muñoz A. The endocrine vitamin D system in the gut. Mol Cell Endocrinol 2017;453:79–87.

[34] He L, Liu T, Shi Y, et al. Gut epithelial vitamin D receptor regulates microbiota-dependent mucosal inflammation by suppressing intestinal epithelial cell apoptosis. Endocrinology 2018;159(2):967–79.

[35] Wang F, Johnson RL, DeSmet ML, Snyder PW, Fairfax KC, Fleet JC. Vitamin D receptor-dependent signaling protects mice from dextran sulfate sodium-induced colitis. Endocrinology 2017;158(6):1951–63.

[36] Wu S, Zhang YG, Lu R, et al. Intestinal epithelial vitamin D receptor deletion leads to defective autophagy in colitis. Gut 2015;64(7):1082–94.

[37] Haifer C, Lawrance IC, Center JR, et al. Vitamin D metabolites are lower with active Crohn's disease and spontaneously recover with development of remission. Ther Adv Gastroenterol 2019;12 1756284819865144.

[38] Meckel K, Li YC, Lim J, et al. Serum 25-hydroxyvitamin D concentration is inversely associated with mucosal inflammation in patients with ulcerative colitis. Am J Clin Nutr 2016;104(1):113–20.

[39] Kim JH, Yamaori S, Tanabe T, et al. Implication of intestinal VDR deficiency in inflammatory bowel disease. Biochim Biophys Acta 2013;1830(1):2118–28.

[40] Cury DB, Oliveira R, Cury MS. Inflammatory bowel diseases: time of diagnosis, environmental factors, clinical course, and management - a follow-up study in a private inflammatory bowel disease center (2003–2017). J Inflamm Res 2019;12:127–35.

[41] Xue LN, Xu KQ, Zhang W, Wang Q, Wu J, Wang XY. Associations between vitamin D receptor polymorphisms and susceptibility to ulcerative colitis and Crohn's disease: a meta-analysis. Inflamm Bowel Dis 2013;19(1):54–60.

[42] Hausmann J, Kubesch A, Amiri M, Filmann N, Blumenstein I. Vitamin D deficiency is associated with increased disease activity in patients with inflammatory bowel disease. J Clin Med 2019;8(9).

[43] Jorgensen SP, Hvas CL, Agnholt J, Christensen LA, Heickendorff L, Dahlerup JF. Active Crohn's disease is associated with low vitamin D levels. J Crohns Colitis 2013;7(10): e407–13.

[44] Lopez-Munoz P, Beltran B, Saez-Gonzalez E, Alba A, Nos P, Iborra M. Influence of Vitamin D deficiency on inflammatory markers and clinical disease activity in IBD patients. Nutrients 2019;11(5).

[45] Raftery T, Merrick M, Healy M, et al. Vitamin D status is associated with intestinal inflammation as measured by fecal calprotectin in Crohn's disease in clinical remission. Dig Dis Sci 2015;60(8):2427–35.

[46] Ryz NR, Lochner A, Bhullar K, et al. Dietary vitamin D3 deficiency alters intestinal mucosal defense and increases susceptibility to *Citrobacter rodentium*-induced colitis. Am J Physiol Gastrointest Liver Physiol 2015;309(9):G730–42.

[47] Schardey J, Globig AM, Janssen C, et al. Vitamin D inhibits pro-Inflammatory T cell function in patients with inflammatory bowel disease. J Crohns Colitis 2019;13 (12):1546–57.

[48] Sharifi A, Vahedi H, Nedjat S, Rafiei H, Hosseinzadeh-Attar MJ. Effect of single-dose injection of vitamin D on immune cytokines in ulcerative colitis patients: a randomized placebo-controlled trial. APMIS 2019;127(10):681–7.

[49] Garg M, Rosella O, Rosella G, Wu Y, Lubel JS, Gibson PR. Evaluation of a 12-week targeted vitamin D supplementation regimen in patients with active inflammatory bowel disease. Clin Nutr 2018;37(4):1375–82.

[50] Narula N, Cooray M, Anglin R, Muqtadir Z, Narula A, Marshall JK. Impact of high-dose vitamin D3 supplementation in patients with Crohn's disease in remission: a pilot randomized double-blind controlled study. Dig Dis Sci 2017;62(2):448–55.

[51] Li J, Chen N, Wang D, Zhang J, Gong X. Efficacy of vitamin D in treatment of inflammatory bowel disease: a meta-analysis. Med (Baltim) 2018;97(46):e12662.

[52] Neu J, Walker WA. Necrotizing enterocolitis. N Engl J Med 2011;364(3):255–64.

[53] Thompson AM, Bizzarro MJ. Necrotizing enterocolitis in newborns: pathogenesis, prevention and management. Drugs 2008;68(9):1227–38.

[54] Shi Y, Liu T, Zhao X, et al. Vitamin D ameliorates neonatal necrotizing enterocolitis via suppressing TLR4 in a murine model. Pediatr Res 2018;83(5):1024–30.

[55] Forssberg H. Microbiome programming of brain development: implications for neurodevelopmental disorders. Dev Med Child Neurol 2019;61(7):744–9.

[56] Liu J, Li Y, Feng Y, et al. Patterned progression of gut microbiota associated with necrotizing enterocolitis and late onset sepsis in preterm infants: a prospective study in a Chinese neonatal intensive care unit. PeerJ 2019;7:e7310.

[57] Jin D, Wu S, Zhang YG, et al. Lack of vitamin D receptor causes dysbiosis and changes the functions of the murine intestinal microbiome. Clin Ther 2015;37(5):996–1009 e1007.

[58] Ooi JH, Li Y, Rogers CJ, Cantorna MT. Vitamin D regulates the gut microbiome and protects mice from dextran sodium sulfate-induced colitis. J Nutr 2013;143:1679–86.

[59] Yoon SS, Sun J. Probiotics, nuclear receptor signaling, and anti-inflammatory pathways. Gastroenterol Res Pract 2011;2011:971938.

[60] Talsness CE, Penders J, Jansen E, Damoiseaux J, Thijs C, Mommers M. Influence of vitamin D on key bacterial taxa in infant microbiota in the KOALA Birth Cohort Study. PLoS One 2017;12(11):e0188011.

[61] Schaffler H, Herlemann DP, Klinitzke P, et al. Vitamin D administration leads to a shift of the intestinal bacterial composition in Crohn's disease patients, but not in healthy controls. J Dig Dis 2018;19(4):225–34.

[62] Kverka M, Zakostelska Z, Klimesova K, et al. Oral administration of *Parabacteroides distasonis* antigens attenuates experimental murine colitis through modulation of immunity and microbiota composition. Clin Exp Immunol 2011;163:250–9.

[63] Dziarski R, Park SY, Des Kashyap R, Dowd SE, Gupta D. *Pglyrp*-regulated gut microflora *Prevotella falsenii*, *Parabacteroides distasonis* and *Bacteroides eggerthii* enhance and *Alistipes finegoldii* attenuates colitis in mice. PLoS One 2016;11:e0146162.

[64] Naderpoor N, Mousa A, Fernanda Gomez Arango L, Barrett HL, Dekker Nitert M, de Courten B. Effect of vitamin D supplementation on faecal microbiota: a randomised clinical trial. Nutrients 2019;11(12).

[65] Waterhouse M, Hope B, Krause L, et al. Vitamin D and the gut microbiome: a systematic review of in vivo studies. Eur J Nutr 2019;58(7):2895–910.

[66] Hooper LV, Littman DR, Macpherson AJ. Interactions between the microbiota and the immune system. Science 2012;336(6086):1268–73.

[67] Mazmanian SK, Liu CH, Tzianabos AO, Kasper DL. An immunomodulatory molecule of symbiotic bacteria directs maturation of the host immune system. Cell 2005;122 (1):107–18.

Chapter 10

Manipulating the gut microbiota

Simone Renwick and Emma Allen-Vercoe
Department of Molecular and Cellular Biology, University of Guelph, Guelph, ON, Canada

Introduction

Technological advances have allowed a broader understanding of the role of the gut microbiota in human health. Culture-independent techniques and metagenomic analyses have provided great insights into the alterations in the composition of the gut microbiota associated with disease. Generally, a "healthy" gut microbiota has a high phylogenetic diversity and stable functional profile [1]. A number of pediatric illnesses, both enteric and systemic, have been linked to an abnormal gut microbiota, although causality has yet to be confirmed.

Antibiotic therapy disrupts the natural ecosystem homeostasis of the gut, and 5%–30% of children experience antibiotic-associated diarrhea (AAD) [2]. Often, AAD is self-limiting, but depleting the microorganisms of the colon can allow invasion and colonization by pathogenic species, such as *Clostridioides difficile*. *C. difficile* infection (CDI) can be life-threatening, and the incidence of infection has been increasing worldwide [3]. Often, CDI can be resolved with an antibiotic regimen. However, 20%–30% of both adult and pediatric patients experience multiple relapses of the infection, termed recurrent (r)CDI [4]. Reoccurrence has been linked to an inability to reestablish normal host-microbiota homeostasis, possibly caused by the administration of broad-spectrum antibiotics that deplete the diversity of gut commensals [5].

Other gastrointestinal (GI) conditions affecting the pediatric population, which are linked to changes in the gut microbiota, include infantile colic, necrotizing enterocolitis (NEC), functional abdominal pain (FAP), inflammatory bowel syndrome (IBS), and inflammatory bowel disease (IBD).

Infantile colic, defined as unexplained fussing or crying for at least 3 hours per day for 3 days or more per week for at least 3 weeks, affects 10%–30% of babies [6]. Crying may be accompanied by flatulence and abdominal distension, but symptoms are usually resolved by 3 months of age. However, it is estimated that 5% of colicky infants have an underlying medical condition, and so all infants need to be assessed [6]. Babies are also at risk of developing the GI disease, NEC, which primarily affects those born prematurely with <1500 g birth

The Developing Microbiome. DOI: https://doi.org/10.1016/B978-0-12-820602-7.00010-6

weight and <32 weeks' gestation [7,8]. NEC is the leading cause of death in infants admitted to the neonatal intensive care unit [7]. Symptoms include abdominal distension, bloody stool, vomiting, and decreased activity; any resulting necrosis can lead to intestinal perforation, sepsis, shock, and death [7]. Although the disease pathogenesis is not well understood, studies have shown that the gut microbiota of NEC patients is abnormally colonized by unexplained blooms of γ-Proteobacteria [8].

High abundances of γ-Proteobacteria are also associated with IBS in children [9]. Symptoms of IBS include abdominal pain, discomfort, and distension, as well as changes in bowel habits. Together, FAP and IBS affect 13.5% of children, 4–18 years old [9]. Although microscopic inflammation has been detected in the colon of some cases, the underlying cause of the disease has yet to be determined [10].

IBD is characterized by chronic inflammation in the gut [11], with the primary manifestations being classified as either Crohn's disease (CD) or ulcerative colitis (UC). CD comprises granulomatous inflammation occurring in patches along the GI tract, whereas UC involves inflammation of the rectum that extends continuously into the colon [11]. Unlike the transmural inflammation of CD, UC does not manifest with granulomas and is limited to the mucosal lining of the colon [11]. IBD affects approximately 4 in every 1000 individuals, with 1 in every 1000 being in the pediatric population (under the age of 18 years old) [11,12]. In addition, 15% of all pediatric IBD cases occur in children (5 years old or younger) [11]. The etiologies of UC and CD are unclear but are likely to involve more complex interactions among the gut microbiota, immune system, environmental factors, and a genetic predisposition [13]. Costly antiinflammatory drugs and immune suppressors, which may have serious side effects, are usually prescribed to induce and maintain remission [14].

Other microbiota-related diseases that extend beyond the colon include obesity, allergic diseases, and autism spectrum disorder (ASD). The increasing prevalence of these diseases in children and adolescents is a major public health concern [15–17]. Obesity is the excessive accumulation of body fat deposits and is the result of an interplay of genetic predisposition, hormones, and environmental factors, in addition to poor diet and a lack of exercise [17]. Overweight and obesity developed during childhood are early risk factors for obesity-related adult morbidity (e.g., hypertension, type 2 diabetes, cardiovascular disease, and some cancer types) and mortality. Colons of obese individuals display chronic, low-grade inflammation, diminished gut barrier function, and an altered gut microbiota [17].

The developed world is experiencing a simultaneous decrease in infectious diseases and increase in allergic diseases and asthma. In early life, the gut microbiota is essential to the education of the developing immune system and the establishment of oral tolerance [18]. A breakdown in these processes may result from the imbalances in the gut microbiota that is observed before the onset of eczema and asthma. There is active research into strategies to prevent allergic diseases, but these are limited by a lack of insight into the underlying mechanisms of allergy development.

Abnormal alterations in the gut microbiota have also been observed in children with ASD, a developmental disorder that affects 1 in every 68 children [16]. Children with ASD experience difficulties with social interactions and communication, and they often display a fixation with repetitive behaviors. The cause of ASD is again unknown, but several environmental factors are being actively investigated, including early exposure to some compounds and viruses, development of autoimmunity, and alterations in the gut microbiota [19]. Children with ASD are also at a higher risk of developing GI disorders that are difficult to treat. Symptoms typically include diarrhea, constipation, and abdominal pain and distension [19]. Diets excluding gluten and casein have been implemented to mitigate symptoms, but with little success [20].

Although it has yet to be established if an abnormal gut microbiota is a cause or consequence of these conditions, manipulation of the gut microbiota has brought some relief of symptoms to patients. We will next review the various methods that are used to modify the gut microbiota in order to treat or prevent disease in the pediatric population—pre, pro-, and synbiotics, as well as fecal microbiota transplantation (FMT) and microbial ecosystem-derived therapeutics.

Prebiotics

According to the Food and Agriculture Organization of the United Nations, prebiotics are nonviable food components that selectively stimulate GI microbiota to confer a health benefit to the host [21]. This definition overlaps with that of dietary fiber. Indeed, most prebiotics are classified as fibers; however, this is not exclusively the case, as polyphenols and lactulose are often also referred to as prebiotics.

Fibrous prebiotics are preferentially utilized by saccharolytic bacteria, including members of the genera *Bifidobacterium* and *Bacteroides*, as well as certain Firmicutes species such as *Faecalibacterium prausnitzii* [22]. Increased abundances of these taxa have been associated with "healthy" microbiomes and have been found to be depleted in many microbiota-related diseased states such as IBD, allergic diseases, and types 1 and 2 diabetes [1].

Naturally occurring prebiotics are found in edible plants, such as fruits and vegetables, whereas artificially produced prebiotics have been developed to supplement foods and beverages, such as infant formula. The most commonly studied prebiotics are lactulose, inulin, fructooligosaccharides (FOS), and galactooligosaccharides (GOS) [23]. Saccharolytic colonic bacteria selectively ferment prebiotics to produce metabolites that are beneficial to the host. For example, short-chain fatty acids (SCFAs) produced by bacteria capable of fermenting prebiotic substrates have a variety of favorable functions in the body, including regulating lipid, glucose, and cholesterol metabolism, promoting mucosal barrier function, nourishing colonic epithelial cells, and stimulating the production of antimicrobial peptides, cytokines, chemokines, and detoxifying enzymes [24]. In addition to modulating the composition and metabolic activity of intestinal microbiota, prebiotics also soften and add bulk to stool [9]. Prebiotics have been studied for their therapeutic properties in a number of pediatric indications, including GI disorders, obesity, allergic diseases, and autism.

Breast milk delivers human milk oligosaccharides (HMOs) to the gut microbiota, complex prebiotics that promote the growth of lactic acid-producing taxa, such as species that belong to the genera *Bifidobacterium* and *Lactobacillus* [25]. Breastfeeding has been associated with healthier long-term outcomes, such as protection against chronic disease (e.g., obesity, cardiovascular disease, and type 2 diabetes) [26]. Whereas the colons of breastfed babies are predominantly colonized by bifidobacteria, those of formula-fed infants tend to display a more adult-like composition, characterized by particular *Bacteroides* and *Clostridium* spp. as well as representatives of the Enterobacteriaceae family of the Proteobacteria phylum, which include opportunistic pathogens and proinflammatory species [25]. Although prebiotics lack the functional diversity of HMOs, fortification of infant formula with inulin, FOS, and GOS has been shown to increase the abundances of health-promoting bifidobacteria to similar levels as those found in breastfed infants [25]. These prebiotics have been shown to be helpful in the treatment of infantile colic; an observational study reported that 79% of colicky infants experienced alleviation of their symptoms after ingesting infant formula supplemented with FOS, GOS, partially hydrolyzed milk proteins, and β-palmitic acid [27]. Subsequent randomized controlled trials (RCTs) revealed a statistically significant reduction in the number of crying episodes in infants that consumed this prebiotic-containing formula compared to infants on a standard formula [28]. Nevertheless, results should be interpreted with caution, as the treatment effects of the prebiotic cannot be discerned from the other components.

At the time of writing, no RCTs have been conducted to study the use of prebiotics for the treatment of FAP and IBS in children, although the efficacy of dietary fiber in general has been studied in several RCTs for these indications in adults, with contradictory results [9]. Three of these studies demonstrated significant relief of symptoms, while the other two showed no difference between treatment groups and control groups. Three RCTs investigating adult IBS found no statistically significant evidence in favor of recommending the use of prebiotics for treatment of symptoms. The etiology of IBS is complex; diets low in fermented oligosaccharides, disaccharides, monosaccharides, and polyols are known to alleviate symptoms of both adult and pediatric IBS. These short-chain carbohydrates, which include prebiotics (such as inulin), are poorly absorbed by the small intestine, and instead retain water and travel to the colon where they undergo rapid fermentation to produce gas; the resulting flatulence and distension can be particularly challenging for

IBS patients [9,29]. It is, therefore, unclear whether prebiotics may be useful in the treatment of IBS symptoms. However, since the gut microbiota differs between individuals, it is difficult to overgeneralize about their potential benefits, as these compounds may be specifically stimulatory to particular microbial taxa present in some individuals but absent in others.

To date, there have been no studies investigating the sole use of prebiotics to treat pediatric IBD and, similarly to IBS, evidence of prebiotic efficacy in alleviating symptoms of adult IBD remains limited. Although administration of prebiotics to adult CD and UC patients has shown promising results for the induction and maintenance of remission [12,30], further studies are needed to better understand the mechanisms behind these reported benefits.

For the treatment of obesity, studies of prebiotic administration to animals and human adults present compelling evidence for beneficial effects including improved control of satiety hormones and glucose metabolism as well as reduced weight and BMI [31]. In overweight or obese children (7–12-years-old), supplementation with oligofructose-enriched inulin (OI) for 16 weeks caused significant decreases in BMI z-scores and percentage body fat, while simultaneously increasing feelings of fullness and increasing secretion of hormones, adiponectin, and ghrelin [31,32]. However, contrasting results were seen in a further RCT for the use of OI in weight management in 7–18-year-old overweight or obese participants, where no changes in BMI or body fat composition were recorded [33].

A meta-analysis and systematic review by Cuello-Garcia et al. [34] investigated the impact of prebiotic consumption on the development of allergic diseases in children. The majority of the 22 RCTs that were analyzed administered FOS or GOS to children under 9 years of age presenting with eczema, asthma/wheezing, self-reported food allergies or "any sign or symptom related to allergy, wheezing, and atopic dermatitis". Evidence suggested that prebiotics may have the potential to reduce the risk of asthma/wheezing, but the quality of the studies were low and had a high risk of bias. As a result, researchers concluded that the uncertain evidence was not of good enough quality to support or reject the potentially beneficial effect of prebiotic supplementation on the development of allergic diseases.

Various dietary interventions have been applied in trials that measure the management of symptoms in pediatric ASD. A prebiotic formulation containing polyphenols (luteolin, quercetin, and rutin) was shown to improve autistic behavior and significantly decrease serum TNFα and IL-6 in children, 4–10 years old [35]. Similarly, positive results (significant improvement in autistic behavior accompanied by reduced GI discomfort) were seen in an RCT involving autistic children, 4–11 years old, where GOS was administered in combination with a diet excluding casein and gluten [20].

In summary, prebiotics appear to have the potential to alter the gut microbiota of pediatric patients to relieve symptoms of some conditions. They are attractive as a treatment modality because they present little to no risk of adverse effects when consumed at an appropriate dose. An additional advantage of prebiotics over probiotics is that prebiotics are less affected by microbial competition in the gut, since they preferentially stimulate beneficial microbes already present in the patients. However, prebiotics lack the ability to induce long-term, sustained effects resulting in the need for continuous dosing.

Probiotics and synbiotics

Probiotics are defined as food products or supplements that contain adequate amounts of live microorganisms that, when consumed, impart a health benefit to the host [36]. Synbiotics are products that contain both probiotic strains and prebiotics, with the expectation that the combination would produce a beneficial, synergistic effect on the host [23]. For the most part, probiotic strains belong to the genera *Lactobacillus* or *Bifidobacterium*, although strains of species representing genera such as *Streptococcus*, *Enterococcus*, *Bacillus*, and *Escherichia* are also the foundations for some probiotic/synbiotic products. Some yeast strains, usually belonging to the *Saccharomyces* genus, may also be utilized as probiotics. Strains are selected based on their safety profile, functionality (i.e., ability to resist stomach acid, bile salts, and digestive enzymes, and to compete with gut microbiota), and manufacturability (i.e., ease of culturing, stability, and resistance to bacteriophages) [23]. In general, it is thought that probiotic strains suppress the growth and activity of potentially pathogenic microorganisms through competition for sites of adhesion and nutrients; and sometimes through the deployment of antimicrobial products, such as acidophiline, bacitracin, and lactacin [23]. Other health-promoting metabolites produced by probiotics include SCFA, amino acids, vitamins, and anticarcinogens, to name a few. Probiotic strains also modulate the immune system by stimulating the production of immunoglobulins and cytokines as well as by promoting the activity of immune cells e.g., macrophages [23]. The therapeutic effect of probiotics on certain disease conditions has been demonstrated through clinical testing.

Probiotics have been actively investigated for the prevention of pediatric AAD and clostridium difficile-associated diarrhea (CDAD). A meta-analysis of five RCTs involving children found *Lactobacillus rhamnosus* GG (LGG) supplements (when given simultaneously with antibiotics) lowered the risk of AAD from 23% in controls to 9.6%, though the optimal dose was not determined [2]. One study found that administration of LGG at a dose of $1-2\times10^{10}$ colony-forming units (CFU) per day lowered the risk by 71% [37]. A probiotic strain of the yeast, *Saccharomyces boulardii*, effectively prevented AAD when administered at $5-40\times10^{9}$ CFU/day [38]. The risk of developing CDAD was lowered in children by 65.9% when *Lactobacillus*, *S. boulardii*, or a mix of probiotic strains were given. Minimal adverse events were reported in all test groups, though the optimal dose was not determined

[39]. Given that probiotic administration generally represents cost-effective and low-risk therapy, this strategy may present a valuable adjunct to the prescription of antibiotics.

Probiotics have commonly been used to target another common pediatric indication: infantile colic. A recent review concluded that there was no clear evidence for the prophylactic use of probiotics for prevention of this indication [6], but, conversely, there is promising evidence for the treatment of infants who have already been diagnosed. Probiotics containing *Lactobacillus reuteri* (strains ATCC 55730 or DSM 17938 at a dose of 10^8 CFU/day) were effective in reducing crying in studies that exclusively enrolled breastfed infants [40]. In many cases, the addition of prebiotics to probiotics, to create synbiotics, bolsters the therapeutic outcomes of probiotics. As well as the beneficial effects of prebiotics discussed earlier, synbiotics may also select specifically for the growth of probiotic strains, thus enhancing their adaptation to the colon. A synbiotic mixture of FOS and seven probiotic strains (*Lactobacillus casei*, *L. rhamnosus*, *Lactobacillus acidophilus*, *Lactobacillus bulgaricus*, *Streptococcus thermophilus*, *Bifidobacterium breve*, and *Bifidobacterium infantis*, each at 10^9 CFU/day) also ameliorated symptoms of colic in breastfed infants [41]. However, evidence for their efficacy in formula-fed infants is limited and does not reflect the same potential. In fact, one RCT reported that administration of *L. reuteri* DSM 17938 at 10^8 CFU/day to formula-fed infants actually increased crying and fussing compared to untreated controls [42].

Multiple meta-analyses and systematic reviews have demonstrated the prophylactic use of probiotic supplementation in reducing the risk of NEC in preterm infants [8]. Data were pooled from studies using products that contained either single strains or mixtures of *B. infantis*, *Bifidobacterium bifidum*, *Bifidobacterium longum*, *Bifidobacterium lactis*, *B. breve*, *S. thermophilus*, *L. acidophilus*, *L. reuteri*, and LGG. The risk of NEC, sepsis, and death was reduced from 6.1%, 14%, and 7.2% to 3.3%, 12%, and 5.1%, respectively, compared to controls [43]. The strongest treatment effects were produced when a dose $>10^9$ CFU/day of each strain was given, and treatment was initiated early in life for 10–14 days, with follow-up courses after each antibiotic treatment [8]. Despite this potential benefit, the safety profile of probiotic use in preterm infants has been called into question, given the immaturity of the intestinal barrier and the associated potential for probiotic-induced sepsis, although adverse events such as these have been rare [8]. In addition, the lack of long-term follow-up studies of premature infants treated with probiotics has limited the ability to predict the lasting effects on these children's health.

Both probiotics and synbiotics have been investigated for their efficacy in treating IBD, but very few RCTs have been conducted in the pediatric population. Probiotics, containing single strains or mixtures of *Escherichia coli*, LGG, *B. longum*, *B. breve*, *B. bifidum*, *Bifidobacterium animalis*, *L. acidophilus*, *Enterococcus faecalis*, *Clostridium butyricum*, or *Bacillus mesentericus* have not been effective for inducing or maintaining remission in adult UC [14].

However, a propriety formulation of eight probiotic strains termed "VSL#3" (which included *Lactobacillus paracasei* DSM 24733, *Lactobacillus plantarum* DSM 24730S, *L. delbrueckii* ssp. *bulgaricus* DSM 24734, *B. longum* DSM 24736, *B. breve* DSM 24732, *B. infantis* DSM 24737, *Streptococcus salivarius* ssp. *thermophilus* DSM24731) was able to induce and maintain remission in pediatric UC patients [44]. VSL#3 (total dose of 1.8×10^{12} CFU/day) was well tolerated and patients who remained enrolled to the end of the 12-month follow-up experienced significant improvements in their symptoms. Clinical improvements were also reported following an RCT involving children with mild to moderate UC limited to the recto-sigmoid area, who received *L. reuteri* enemas [45].

Of the few studies of probiotic intervention in children with CD, insufficient evidence was found to support the use of preparations containing *E. coli* or LGG [45]. Results of adult CD studies are similarly disappointing, with most trials again finding little ability of probiotics (LGG, *Lactobacillus johnsonii*, or VSL#3) to induce or maintain remission [12,14]. The inconsistency of these results is likely related to the differences in the probiotic strains and species used. Thus, further studies are warranted to examine preparations that have seemed to show benefits in the IBD patient population.

As well as probiotics, synbiotic preparations have been studied for their effectiveness in the treatment of IBD. A preparation of an OI prebiotic combined with an unspecified strain of *B. longum*, together termed "Synergy 1," was found to be mildly beneficial in improving CD and UC clinical symptoms in adult patients [46,47]. Similarly, a combination of strains of *B. breve*, *B. longum*, and *L. casei* (total dose of 2.25×10^{11} CFU/day) with psyllium fiber significantly improved remission rates in adult UC patients [48]. In contrast, a mixture of representative strains of *Pediococcus pentoseceus*, *Lactococcus raffinolactis*, *L. paracasei* ssp. *paracasei*, and *L. plantarum* with four fermentable fibers (β-glucans, inulin, pectin, and resistant starch) did not induce any lasting remission [49].

The aforementioned VSL#3 supplement was shown to be effective in treating children and teenagers with IBS, significantly reducing abdominal pain/discomfort, bloating, and flatulence. Probiotic products containing LGG and *L. reuteri* strains also ameliorated IBS symptoms in children, for up to 4 weeks after ending treatment. However, RCTs have not demonstrated any improvement in diarrhea or constipation in these children [45]. A 2016 RCT compared the efficacy of a probiotic (5×10^9 CFU/day *B. lactis* B94), a prebiotic inulin supplement, and a synbiotic combination of these in 71 children, 4–16 years old [10]. The patients administered the prebiotic experienced no clear, significant improvements in their symptoms. The probiotic group experienced significant relief of bloating, belching-abdominal distension, and difficulty defecating, whereas the synbiotic group experienced the same, in addition to improved mucoid stool. Rates of full recovery for the synbiotic, probiotic, and prebiotic groups were 39.1%, 29.2%, and 12.5%, respectively. No adverse effects were reported by any of these participants.

A review of the use of probiotics for the treatment of overweight and obesity considered data from five RCTs involving children and adolescents [17]. Across these studies, there were generally positive impacts on weight gain and waist circumference. One study revealed improved insulin resistance and another affirmed decreased hypertransaminasemia. Supplements included either individual strains of LGG or *Bifidobacterium pseudocatenulatum* CECT 7765, or mixtures of *L. acidophilus* ATCC B3208, *L. rhamnosus* DSMZ 21690, *B. lactis* DSMZ 32,296, *B. bifidum* ATCC SD6576, or VSL#3, with all strains administered $\geq 10^9$ CFU/day. An interesting study by Luoto et al. [50] administered either a placebo or a probiotic containing LGG to pregnant women (10^{10} CFU/day) for 4 weeks before delivery, then to the mother while breastfeeding or directly to weaned infants for 6 months. Infants were then followed for a further 10 years revealing that their infantile prebiotic treatment did not significantly prevent or reduce weight gain in the long term.

Probiotic preparations have been shown to be useful for preventing eczema in infants in multiple meta-analyses and systematic reviews [15,51]. Pooled analysis of 4755 children found fewer cases of eczema (28.22%) compared to control (35.67%) when probiotics were administered before 3 months of age [51]. In terms of the strain used, the studies were highly heterogeneous, but subgroup analysis found the strongest treatment effect when a mixture of strains were administered. Species of lactobacilli and bifidobacteria were not significantly useful in preventing eczema. The success for eczema was not reflected in other allergic conditions, such as food allergies, asthma/wheezing, and allergic rhinitis, since the evidence does not support a prophylactic use [18].

Probiotic preparations have been assessed for their abilities to alleviate symptoms of ASD; however, the results of these RCTs were inconsistent. Several studies reported improvements in GI complaints and ASD behavior when supplemented with 10^{10} CFU/day *L. acidophilus* Rosell-11; a mixture of 5×10^9 CFU/day of strains of *L. acidophilus*, *L. rhamnosus*, and *B. longum*; or the commercial product, Delpro (6×10^9 CFU/day each of unspecified strains of *L. acidophilus*, *L. casei*, *L. delbruecki*, *B. longum*, and *B. bifidum*) [52–54]. In contrast, children who received supplements of 6×10^9 CFU/day of *L. rhamnosus* HN001 or 9×10^9 CFU/day of *B. animalis* ssp. *lactis* HN019 for the first 2 years of life displayed no differences in terms of mood, behavior, or cognition compared to controls during a 10-year follow-up [55].

Probiotics and synbiotics have been investigated for a wide range of conditions, including several pediatric illnesses. Most products have been well tolerated in children (0–18 years old), and the majority of trials have reported few to no adverse effects [56]. Clinical evidence supports the therapeutic role of certain probiotic strains and combinations for some diseases, but efficacy may depend on the strain/cultures and dose administered. Most systematic reviews and meta-analyses discussed here drew conclusions from data pooled from studies that used different strains/combinations, dosages, study durations, and outcomes assessed, which make general conclusions difficult to obtain.

Probiotics are regulated as food supplements by agencies such as the US Food and Drug Administration, and as such the standards for their manufacture and quality testing are much lower than for drug products [57]. Investigations of commercial probiotics have found inconsistencies between product labels and true product composition. For example, strains listed on product labels were absent or reduced to ineffective numbers, and contaminating strains were recovered from products which, in some cases, included pathogens [57]. It is clear that more stringent, standardized, and validated quality control procedures for commercialized probiotic products are warranted. Overall, administration of pre-, pro-, and synbiotics to preterm infants and other immunocompromised patients should be carefully considered with a thorough risk-benefit analysis [38].

Fecal microbiota transplantation

The use of fecal matter to treat disease was documented as early as 4th century China; for example, severe food poisoning, typhoid fever, and diarrhea were remedied by oral administrations of fecal suspensions [58]. This practice has evolved into the FMT technique that is practiced today. FMT is defined as the transfer of the microbiota from a healthy donor's stool to the colon of a patient experiencing disease as a result of an abnormal gut microbial ecosystem [59]. The aim of this practice is to introduce strains of a healthy donor's ecosystem in order to colonize the GI tract of the recipient, either combining with or replacing the recipient's damaged or altered microbiota. Fecal suspensions can be delivered via the upper GI tract (i.e., through nasogastric, nasoduodenal, or nasojejunal tubes) or via the lower GI tract (i.e., through colonoscopy or rectal enemas) [59].

Because of the increasing incidence of life-threatening rCDI, FMT is being actively investigated as an alternative to protracted oral vancomycin instillation. The first Western record of FMT use was for the successful treatment of rCDI (documented as pseudomembranous colitis) in 1958 [60]. Subsequently, in 1983, FMT instilled via enema cured a 65-year-old woman with severe rCDI [61]. Since then, there have been numerous reports (including retrospective studies, uncontrolled prospective cohorts, and RCTs) that have demonstrated the ability of FMT to clinically cure rCDI in adults, with a mean success rate of 90% and long-term remission lasting at least 2 years [62,63]. As a result, the therapy is now strongly recommended in the clinical practice guidelines for adult patients experiencing their second or subsequent relapse after a pulsed vancomycin regimen [3].

At the time of writing, no RCTs have evaluated the efficacy and safety of FMT for pediatric rCDI. Based on the therapeutic success of FMT for adult rCDI, the clinical guidelines weakly recommend the use of FMT for pediatric rCDI, with oral vancomycin treatment being the primary recommendation [3]. Thus far, several small case studies have enrolled pediatric patients to receive FMT, reporting a high overall success rate of cure (mostly 100%), regardless of mode of delivery (nasogastric, nasojejunal, or colonoscopy were used) [64–66]. Adverse events included self-limiting abdominal pain, vomiting, and mucoid stool. Follow-up periods have been variable, but children followed for 2 years have reported no long-term complications arising from the procedure [66]. Fecal samples collected approximately 6 months post-FMT and analyzed for microbiome composition displayed increased alpha diversity, a sevenfold increase in the ratio of Bacteroidetes to Firmicutes, and a taxonomic profile that resembled the donor stool [65]. The largest retrospective cohort involving pediatric patients published to date found that FMT resolved rCDI in 86.6% of the 335 patients enrolled (aged 11 months to 23 years) with no relapses during the 2 months of follow-up [59]. Despite the success reported in several cohorts, RCTs are still needed to better assess the efficacy of FMT in comparison to suitable controls. At the time of writing, several RCTs are listed as either recruiting or completed on ClinicalTrials.gov, but conclusive results have yet to be published.

The therapeutic success of FMT for rCDI has prompted investigations into its clinical applications to other gut microbiota-related diseases, such as IBD. Paramsothy et al. [67] performed a systematic review and meta-analysis on the case studies, cohorts, and RCTs in which FMT was used to treat IBD and found clinical remission in 36% of 555 UC patients and 50.5% of 83 CD patients. Subgroup analysis found that remission was achieved in 23% of pediatric UC and 54% of pediatric CD patients. For pediatric patients, follow-up time varied widely, ranging from 4 weeks to 10 months, with no standardization across studies for data collection. Minor adverse reactions included diarrhea, abdominal distension and pain, fever, and flatulence following FMT [67]. Increased microbial diversity was observed accompanied by a shift in taxonomic profile toward that of the donor, which persisted for 8 weeks post-FMT. However, the data available suggested that patients with IBD would invariably experience a relapse. As such, researchers called for investigations into the use of FMT for maintenance therapy. An updated meta-analysis by Fang et al. [68] found a remission rate of 10% for pediatric UC (26% in adults) and 45% for pediatric CD (22% for adults). This study also found that, of persons with mild–moderate IBD, 20.9% experienced improved outcomes while 32.3% of moderate–severe IBD patients achieved remission, indicating that FMT may be more effective for the treatment of severe IBD than mild–moderate IBD [68,69].

Despite the challenges of treating the developing microbiome, which may account for the variation in results between adult and pediatric studies, FMT presents a promising avenue in the treatment of pediatric IBD and multiple

RCTs are in progress (ClinicalTrials.gov). Nevertheless, rates of remission for FMT-treated IBD patients fall short in comparison to the cure rate of rCDI, likely because IBD is a more complex disease than CDI. CDI results from a decline in the phylogenetic diversity of intestinal microbiota, often the result of antibiotic use, and FMT is efficacious in replenishing the diversity lost. In contrast, the etiologies of UC and CD are unclear but are known to involve more complex interactions among the gut microbiota, immune system, environmental factors and a genetic predisposition [13]. Hence, consideration should also be given to the risk of FMT triggering or worsening IBD inflammation. A meta-analysis of case studies, cohort studies, and RCTs by Qazi et al. [70] found worsening IBD symptoms in 14.3% of patients after FMT treatment. Subgroup analysis of only RCTs (since a control group allows distinction between IBD-related flares and FMT-provoked flares) found that a marginal 4.6% of adult patients experienced worsening IBD activity (pediatric data not available). As such, potential benefits of FMT to IBD may outweigh the risk of increased flares.

The therapeutic use of FMT is being considered for other diseases associated with an altered intestinal microbiota, such as IBS, constipation, metabolic diseases, neuropsychiatric disorders, autoimmune disorders, and allergic disorders [4,71,72]. Although these studies show promising results for adult patients, evidence for the pediatric population is very limited. Case studies involving single pediatric patients have shown improvements for children suffering from Tourette's syndrome and ASD for up to 8 weeks post-FMT before symptoms relapsed [73,74]. As a result, there are ongoing pediatric clinical trials for ASD and Tourette's syndrome on ClinicalTrials.gov. In addition, a number of other RCTs have been registered for pediatric disease indications, including mesothelioma, hematopoietic and lymphoid cell neoplasm, acute myeloid lymphoma, acute graft versus host disease, and severe acute malnutrition.

Compared to treatment with pre-, pro- and synbiotics, FMT is able to abruptly replenish the diminished diversity of patients' gut microbiota without continuous administration. Generally, FMT procedures have been well tolerated by patients and minor, self-limiting adverse effects have included vomiting, transient fever, abdominal pain, cramping, distension, flatulence, nausea, bloody stool, constipation, and diarrhea [4,59,67,75]. Administration of FMT via the upper GI tract has also resulted in nasal congestion and sore throat, which lasted less than 24 hours. Severe adverse events have included infections, IBD flares, colitis, and, rarely, death due to sepsis or aspiration of fecal material during upper GI administration [4]. Severe adverse reactions were estimated to occur in 2.5% of pediatric patients based on the studies available [59]. Lastly, patients were more likely to develop adverse effects after receiving FMT via the upper GI tract compared to the lower GI tract [76].

Although studies have demonstrated the safety and efficacy of FMT in immunocompromised adults [59], the risk of transferring an undetected pathogenic agent from donor to recipient is very real. Post-FMT bacterial and viral infections have been reported [76]. As such, careful donor screening plays an important role in minimizing the risk of the procedure. Potential donors are typically screened for blood-borne and enteric pathogens, GI disorders, chronic diseases, and high-risk behaviors [77]. Surprisingly, a large number of asymptomatic donors were found to carry parasites. As a result of these strict criteria, over 90% of potential donors fail screening evaluations, presenting a major challenge to FMT administration [78].

Shifts in the viral and metabolomic profiles of the recipient's microbiota toward that of the donor's were observed 4 weeks post-FMT, and compositional changes were still detected 6 months post-FMT [65,69]. Since most pediatric patients receive stool from an adult donor, commonly a close relative, concerns are raised regarding the impact of transplanting an adult microbiota into the colon of a child. Typically, studies have had short follow-up periods, limiting assessment of the long-term impact and sustained effects on the developing microbiome.

Ecosystem-derived therapeutics

Standardized mixtures of bacterial strains derived from human stool are being developed into therapeutics for the treatment of gut microbiota-related diseases [5]. This approach, termed by some as "microbial ecosystem therapeutics" (METs), uses collections of microbes described as "synthetic stool" (selected strains form a simplified microbial community and are not necessarily probiotic strains). METs are expected to represent an improvement on standard FMT as they aim to combine the effectiveness of FMT therapy in restoring a damaged intestinal microbiota while mitigating some of the FMT-related safety concerns and challenges [79]. Only pure and defined cultures of selected bacterial strains are combined to create MET products, minimizing the risk of transferring pathogenic agents to recipients. The therapeutic formulations also have the distinct benefit of not relying on a continuous supply of fecal material from donors, thus avoiding the challenges that arise in FMT donor recruitment. Several companies have encapsulated their METs to allow oral delivery, removing the risks of anesthesia or damage to the GI tract faced during FMT procedures that use invasive instillation techniques such as colonoscopy.

The first MET was tested by Tvede and Rask-Madsen in 1989 for the treatment of rCDI in five adult patients [80]. A mixture of 10 bacterial strains derived from stool were introduced to the patients via rectal enemas. All patients experienced alleviation of their symptoms within 24 hours and did not experience any relapses up to the 4 months of follow-up provided. However, METs were not further investigated as a therapy because CDI was uncommon at the time, and usually effectively treated using antibiotic therapies. Over two decades later, Petrof et al. [5] demonstrated the efficacy of a MET in the successful treatment of two adult rCDI patients in a small proof-of-principle study. A consortium of 33 strains, termed "RePOOPulate," was inoculated into patients via colonoscopy and resolved their infections, with no relapses noted during the subsequent 6 months of follow-up. A subset of taxa representative of the instilled strains was shown to colonize the patients through to the end of the study, as determined through assessments of fecal samples taken during the follow-up period.

In the past decade, several biotechnology companies have begun exploring the market for ecosystem-derived therapeutics for the manipulation of the gut microbiota [81]. Seres Therapeutics Inc. cured 86.7% of 30 adult rCDI patients with a preparation of endospores fractionated from adult stool samples, termed "SER-109" [82]. Of these patients, no relapses were reported during the 8 weeks of follow-up. Although SER-109 was subsequently not proven to be efficacious in a phase II RCT (44% of the treatment group experienced a relapse, compared to 53% of controls) [83], at time of writing, a phase III RCT is currently ongoing (ClinicalTrials.gov Identifier: NCT03183128).

Several other companies have recently reported results for their MET preparations, and at the time of writing, there are numerous trials ongoing. Rebiotix Inc. derived a microbial suspension from the stool of four donors, termed "RBX2660," which effectively cured rCDI in 87.1% of adult participants up to a 6-month follow-up in a phase II trial (ClinicalTrials.gov Identifier: NCT01925417) [84]. Vedanta Biosciences Inc. has reported the safety of their product VE303, a mixture of eight commensal strains, for the prevention of CDI, and a phase II RCT is ongoing (ClinicalTrials.gov Identifier: NCT03788434). Other prophylactic products that target the microbiota to prevent CDI include SYN-004 by Synthetic Biologics Inc. (a β-lactam antibiotic degrading enzyme) [85] and PlyCD (a prophage lysin that specifically targets *C. difficile* in the gut) [86]. Thus far, METs have not been tested for CDI in a pediatric population.

Beyond treating rCDI, a handful of MET products are in the early stages of testing safety and tolerability in phase I RCTs involving adult patients. Seres Therapeutics Inc. has developed SER-287, produced in the same manner as SER-109, for the treatment of adult UC. The product was reported to be safe, and a phase II RCT is currently testing its efficacy (ClinicalTrials.gov Identifier: NCT03759041). In addition, Vedanta Biosciences Inc. is currently testing the defined bacterial consortium, VE416, in adults and children (≥12 years old) for the treatment of peanut allergies (ClinicalTrials.gov Identifier: NCT03936998). NuBiyota LLC. has developed multiple METs similar to the "RePOOPulate" prototype consortium and is currently testing them in several separate RCTs in a number of distinct indications; for the treatment of depression and anxiety, solid tumors, head and neck squamous cell carcinoma, UC, obesity, and CDI in adults (ClinicalTrials.gov). Results from these trials are not yet available.

Little can be deduced regarding the impact of the nascent MET principle on pediatric diseases without further research into the safety and efficacy of such products on the adult population. Similar to FMT, there are many factors that need to be carefully considered beforehand, including how transferring bacteria isolated from an adult to an infant will impact the developing microbiome. A healthy gut microbiota has yet to be clearly defined, posing a challenge for selection of the most efficacious strains. The strains that comprised the RePOOPulate formulation, for example, were chosen not only because they were representative of taxa commonly found in the healthy human gut, but also based on their ability to be cultured and their susceptibility to antimicrobials. As these taxa were shown to persist in subjects for up to 6 months postprocedure, it is important to understand whether MET-introduced strains might have long-term impacts on factors such as development in the pediatric population. Some may argue that the benefits of curing severe diseases outweigh the risks involved in MET administration, but robust long-term follow-up studies are required before any conclusions can be drawn. MET also faces regulatory challenges as products that fall within this group are classified as both biologic agents and manufactured products, generally requiring a much higher standard to be met than for FMT preparations or probiotics [79]. As such, bringing MET products to market may be a slow process despite the intense interest, although in the long term, much can be learned from the results of MET treatments that use standardized formulations as their basis.

Conclusion

Our expanding knowledge of the role of the gut microbiota in human health and disease, along with significant advances in technology, has resulted in the development of several techniques for the therapeutic manipulation of the microbial ecosystem. Pre-, pro-, and synbiotics have excellent safety profiles; and based on the clinical evidence available, these products appear to bring some relief to pediatric patients when administered at the right doses; and for pro- and synbiotics, when the right cultures are used. However, there is a lack of evidence to support any long-term, sustained effects, resulting in the need for continuous dosing. It is also clear that more stringent, standardized, and validated quality control procedures are needed for commercialized probiotics, due to the inconsistencies between product labels and true product composition.

The complete transfer of a "healthy" gut ecosystem to the colon of a patient using FMT has been promising for the treatment of pediatric rCDI and pediatric IBD. FMT also appears to have the potential to induce long-term, sustained effects as changes in recipients' microbiomes were detected up to 6 months post-FMT. Still, the evidence for the therapeutic use of FMT for other disease indications is limited due to a lack of RCTs. In addition, severe adverse reactions to FMTs have been recorded in 2.5% of pediatric recipients, indicating the need for a therapeutic product that is both extremely safe and highly effective. Born from this necessity are ecosystem-based therapeutics, such as METs. However, these products are still in their infancy as most are currently in the early stages of testing their safety and tolerability in adult clinical trials. Therefore, it may be some time before such preparations are readily available for the treatment of the pediatric population.

Funding sources

Canadian Institutes of Health Research, National Institutes of Health

Disclosure statement

EA-V is CSFO of NuBiyota LLC, a company developing microbial ecosystem therapeutics for pharmaceutical use.

References

[1] Lloyd-Price J, Abu-Ali G, Huttenhower C. The healthy human microbiome. Genome Med 2016;8(1):51.

[2] Szajewska H, Kolodziej M. Systematic review with meta-analysis: *Lactobacillus rhamnosus* GG in the prevention of antibiotic-associated diarrhoea in children and adults. Aliment Pharmacol Ther 2015;42(10):1149–57.

[3] Mcdonald LC, Gerding DN, Johnson S, Bakken JS, Carroll KC, Coffin SE, et al. Clinical practice guidelines for *Clostridium difficile* infection in adults and children: 2017 update by the Infectious Diseases Society of America (IDSA) and Society for Healthcare Epidemiology of America (SHEA). Clin Infect Dis 2018;66(7):e1–e48.

[4] Davidovics ZH, Michail S, Nicholson MR, Kociolek LK, Pai N, Hansen R, et al. Fecal microbiota transplantation for recurrent *Clostridium difficile* infection and other conditions in children: a joint position paper from the North American Society for Pediatric Gastroenterology, Hepatology, and Nutrition and the European Society for Pediatric Gastroenterology, Hepatology, and Nutrition. J Pediatr Gastroenterol Nutr 2019;68(1):130–43.

[5] Petrof EO, Gloor GB, Vanner SJ, Weese SJ, Carter D, Daigneault MC, et al. Stool substitute transplant therapy for the eradication of *Clostridium difficile* infection: 'RePOOPulating' the gut. Microbiome 2013;1(1):3.

[6] Ong TG, Gordon M, Banks SS, Thomas MR, Akobeng AK. Probiotics to prevent infantile colic. Cochrane Database Syst Rev 2019;3:CD012473.

[7] Bellodas Sanchez J, Kadrofske M. Necrotizing enterocolitis. Neurogastroenterol Motil 2019;31(3):e13569.

[8] Patel RM, Underwood MA. Probiotics and necrotizing enterocolitis. Semin Pediatr Surg 2018;27(1):39–46.

[9] Wegh CAM, Schoterman MHC, Vaughan EE, Belzer C, Benninga MA. The effect of fiber and prebiotics on children's gastrointestinal disorders and microbiome. Expert Rev Gastroenterol Hepatol 2017;11(11):1031–45.

[10] Basturk A, Artan R, Yilmaz A. Efficacy of synbiotic, probiotic, and prebiotic treatments for irritable bowel syndrome in children: a randomized controlled trial. Turk J Gastroenterol 2016;27(5):439–43.

[11] Conrad MA, Rosh JR. Pediatric inflammatory bowel disease. Pediatr Clin North Am 2017;64(3):577–91.

[12] Hansen JJ, Sartor RB. Therapeutic manipulation of the microbiome in IBD: current results and future approaches. Curr Treat Options Gastroenterol 2015;13(1):105–20.

[13] Bouma G, Strober W. The immunological and genetic basis of inflammatory bowel disease. Nat Rev Immunol 2003;3(7):521–33.

[14] Derwa Y, Gracie DJ, Hamlin PJ, Ford AC. Systematic review with meta-analysis: the efficacy of probiotics in inflammatory bowel disease. Aliment Pharmacol Ther 2017;46(4):389–400.

[15] Cuello-Garcia CA, Brozek JL, Fiocchi A, Pawankar R, Yepes-Nunez JJ, Terracciano L, et al. Probiotics for the prevention of allergy: a systematic review and meta-analysis of randomized controlled trials. J Allergy Clin Immunol 2015;136(4):952–61.

[16] Van De Sande MMH, Van Buul VJ, Brouns JPH. Autism and nutrition: the role of the gut-brain axis. Nat Res Rev 2014;27(2):199–214.

[17] Cerdo T, Garcia-Santos JA, Bermúdez MG, Campoy C. The role of probiotics and prebiotics in the prevention and treatment of obesity. Nutrients 2019;11(3).

[18] West CE, Dzidic M, Prescott SL, Jenmalm MC. Bugging allergy; role of pre-, pro- and synbiotics in allergy prevention. Allergol Int 2017;66(4):529–38.

[19] Ng QX, Loke W, Venkatanarayanan N, Lim DY, Soh AYS, Yeo WS. A systematic review of the role of prebiotics and probiotics in autism spectrum disorders. Medicina 2019;55(5).

[20] Grimaldi R, Gibson GR, Vulevic J, Giallourou N, Castro-Mejia JL, Hansen LH, et al. A prebiotic intervention study in children with autism spectrum disorders (ASDs). Microbiome 2018;6(1):133.

[21] Pineiro M, Asp NG, Reid G, Macfarlane S, Morelli L, Brunser O, et al. FAO technical meeting on prebiotics. J Clin Gastroenterol 2008;42(Suppl. 3 Pt. 2):S156–9.

[22] Oliphant K, Allen-Vercoe E. Macronutrient metabolism by the human gut microbiome: major fermentation by-products and their impact on host health. Microbiome 2019;7(1):91.

[23] Markowiak P, Slizewska K. Effects of probiotics, prebiotics, and synbiotics on human health. Nutrients 2017;9(9).

[24] Fernández J, Redondo-Blanco S, Gutiérrez-Del-Río I, Miguélez EM, Villar CJ, Lombó F. Colon microbiota fermentation of dietary prebiotics towards short-chain fatty acids and their roles as anti-inflammatory and antitumour agents: a review. J Funct Foods 2016;25:511–22.

[25] Penders J. Early diet and the infant gut microbiome: how breastfeeding and solid foods shape the microbiome. In: Browne PD, Claassen E, Cabana M, editors. Microbiota in health and disease: from pregnancy to childhood. Wageningen, The Netherlands: Wageningen Academic Publishers; 2017. p. 105–18.

[26] Binns C, Lee M, Low WY. The long-term public health benefits of breastfeeding. Asia Pac J Public Health 2016;28(1):7–14.

[27] Savino F, Cresi F, Maccario S, Cavallo F, Dalmasso P, Fanaro S, et al. "Minor" feeding problems during the first months of life: effect of a partially hydrolysed milk formula containing fructo- and galacto-oligosaccharides. Acta Paediatr Suppl 2003;91(441):86–90.

[28] Savino F, Palumeri E, Castagno E, Cresi F, Dalmasso P, Cavallo F, et al. Reduction of crying episodes owing to infantile colic: a randomized controlled study on the efficacy of a new infant formula. Eur J Clin Nutr 2006;60(11):1304–10.

[29] Ooi SL, Correa D, Pak SC. Probiotics, prebiotics, and low FODMAP diet for irritable bowel syndrome – what is the current evidence? Complement Ther Med 2019;43:73–80.

[30] Derikx LA, Dieleman LA, Hoentjen F. Probiotics and prebiotics in ulcerative colitis. Best Pract Res Clin Gastroenterol 2016;30(1):55–71.

[31] Nicolucci AC, Reimer RA. Prebiotics as a modulator of gut microbiota in paediatric obesity. Pediatr Obes 2017;12(4):265–73.

[32] Nicolucci AC, Hume MP, Martinez I, Mayengbam S, Walter J, Reimer RA. Prebiotics reduce body fat and alter intestinal microbiota in children who are overweight or with obesity. Gastroenterology 2017;153(3):711–22.

[33] Liber A, Szajewska H. Effect of oligofructose supplementation on body weight in overweight and obese children: a randomised, double-blind, placebo-controlled trial. Br J Nutr 2014;112(12):2068–74.

[34] Cuello-Garcia C, Fiocchi A, Pawankar R, Yepes-Nunez JJ, Morgano GP, Zhang Y, et al. Prebiotics for the prevention of allergies: a systematic review and meta-analysis of randomized controlled trials. Clin Exp Allergy 2017;47(11):1468–77.

[35] Tsilioni I, Taliou A, Francis K, Theoharides TC. Children with autism spectrum disorders, who improved with a luteolin-containing dietary formulation, show reduced serum levels of TNF and IL-6. Transl Psychiatry 2015;5:e647.

[36] Hill C, Guarner F, Reid G, Gibson GR, Merenstein DJ, Pot B, et al. Expert consensus document. The International Scientific Association for Probiotics and Prebiotics consensus statement on the scope and appropriate use of the term probiotic. Nat Rev Gastroenterol Hepatol 2014;11(8):506–14.

[37] Vanderhoof JA, Whitney DB, Antonson DL, Hanner TL, Lupo JV, Young RJ. *Lactobacillus* GG in the prevention of antibiotic-associated diarrhea in children. J Pediatr 1999;135(5):564–8.

[38] Goldenberg JZ, Lytvyn L, Steurich J, Parkin P, Mahant S, Johnston BC. Probiotics for the prevention of pediatric antibiotic-associated diarrhea. Cochrane Database Syst Rev 2015;(12):CD004827.

[39] Lau CS, Chamberlain RS. Probiotics are effective at preventing *Clostridium difficile*-associated diarrhea: a systematic review and meta-analysis. Int J Gen Med 2016;9:27–37.

[40] Barnes D, Yeh AM. Bugs and guts: practical applications of probiotics for gastrointestinal disorders in children. Nutr Clin Pract 2015;30(6):747–59.

[41] Kianifar H, Ahanchian H, Grover Z, Jafari S, Noorbakhsh Z, Khakshour A, et al. Synbiotic in the management of infantile colic: a randomised controlled trial. J Paediatr Child Health 2014;50(10):801–5.

[42] Sung V, Hiscock H, Tang ML, Mensah FK, Nation ML, Satzke C, et al. Treating infant colic with the probiotic *Lactobacillus reuteri*: double blind, placebo controlled randomised trial. BMJ 2014;348:g2107.

[43] Underwood MA. Impact of probiotics on necrotizing enterocolitis. Semin Perinatol 2017;41(1):41–51.

[44] Fedorak RN, Feagan BG, Hotte N, Leddin D, Dieleman LA, Petrunia DM, et al. The probiotic VSL#3 has anti-inflammatory effects and could reduce endoscopic recurrence after surgery for Crohn's disease. Clin Gastroenterol Hepatol 2015;13(5):928–35 e2.

[45] Guandalini S. Are probiotics or prebiotics useful in pediatric irritable bowel syndrome or inflammatory bowel disease? Front Med 2014;1:23.

[46] Furrie E, Macfarlane S, Kennedy A, Cummings JH, Walsh SV, O'neil DA, et al. Synbiotic therapy (*Bifidobacterium longum*/Synergy 1) initiates resolution of inflammation in patients with active ulcerative colitis: a randomised controlled pilot trial. Gut 2005;54(2):242–9.

[47] Steed H, Macfarlane GT, Blackett KL, Bahrami B, Reynolds N, Walsh SV, et al. Clinical trial: the microbiological and immunological effects of synbiotic consumption - a randomized double-blind placebo-controlled study in active Crohn's disease. Aliment Pharmacol Ther 2010;32(7):872–83.

[48] Fujimori S, Tatsuguchi A, Gudis K, Kishida T, Mitsui K, Ehara A, et al. High dose probiotic and prebiotic cotherapy for remission induction of active Crohn's disease. J Gastroenterol Hepatol 2007;22(8):1199–204.

[49] Chermesh I, Tamir A, Reshef R, Chowers Y, Suissa A, Katz D, et al. Failure of Synbiotic 2000 to prevent postoperative recurrence of Crohn's disease. Dig Dis Sci 2007;52 (2):385–9.

[50] Luoto R, Kalliomaki M, Laitinen K, Isolauri E. The impact of perinatal probiotic intervention on the development of overweight and obesity: follow-up study from birth to 10 years. Int J Obes 2010;34(10):1531–7.

[51] Zuccotti G, Meneghin F, Aceti A, Barone G, Callegari ML, Di Mauro A, et al. Probiotics for prevention of atopic diseases in infants: systematic review and meta-analysis. Allergy 2015;70(11):1356–71.

[52] Kaluzna-Czaplinska J, Blaszczyk S. The level of arabinitol in autistic children after probiotic therapy. Nutrition 2012;28(2):124–6.

[53] Shaaban SY, El Gendy YG, Mehanna NS, El-Senousy WM, El-Feki HSA, Saad K, et al. The role of probiotics in children with autism spectrum disorder: a prospective, open-label study. Nutr Neurosci 2018;21(9):676–81.

[54] West R, Roberts E, Sichel LS, Sichel J. Improvements in gastrointestinal symptoms among children with autism spectrum disorder receiving the Delpro® probiotic and immunomodulator formulation. J Prob Health 2013;1(1).

[55] Slykerman RF, Kang J, Van Zyl N, Barthow C, Wickens K, Stanley T, et al. Effect of early probiotic supplementation on childhood cognition, behaviour and mood a randomised, placebo-controlled trial. Acta Paediatr 2018;107(12):2172–8.

[56] Van Den Nieuwboer M, Brummer RJ, Guarner F, Morelli L, Cabana M, Claassen E. Safety of probiotics and synbiotics in children under 18 years of age. Benef Microbes 2015;6(5):615–30.

[57] Kolacek S, Hojsak I, Berni Canani R, Guarino A, Indrio F, Orel R, et al. Commercial probiotic products: a call for improved quality control. A position paper by the ESPGHAN Working Group for probiotics and prebiotics. J Pediatr Gastroenterol Nutr 2017;65 (1):117–24.

[58] Zhang F, Cui B, He X, Nie Y, Wu K, Fan D, et al. Microbiota transplantation: concept, methodology and strategy for its modernization. Protein Cell 2018;9(5):462–73.

[59] Nicholson MR, Mitchell PD, Alexander E, Ballal S, Bartlett M, Becker P, et al. Efficacy of fecal microbiota transplantation for *Clostridium difficile* infection in children. Clin Gastroenterol Hepatol 2019;.

[60] Eiseman B, Silen W, Bascom GS, Kauvar AJ. Fecal enema as an adjunct in the treatment of pseudomembranous enterocolitis. Surgery 1958;44(5):854–9.

[61] Schwan A, Sjolin S, Trottestam U, Aronsson B. Relapsing *Clostridium difficile* enterocolitis cured by rectal infusion of homologous faeces. Lancet 1983;2(8354):845.

[62] Lee CH, Belanger JE, Kassam Z, Smieja M, Higgins D, Broukhanski G, et al. The outcome and long-term follow-up of 94 patients with recurrent and refractory *Clostridium difficile* infection using single to multiple fecal microbiota transplantation via retention enema. Eur J Clin Microbiol Infect Dis 2014;33(8):1425–8.

[63] Surawicz CM, Brandt LJ, Binion DG, Ananthakrishnan AN, Curry SR, Gilligan PH, et al. Guidelines for diagnosis, treatment, and prevention of *Clostridium difficile* infections. Am J Gastroenterol 2013;108(4):478–98 quiz 499.

[64] Barfield E, Small L, Navallo L, Solomon A. Going to the bank: fecal microbiota transplantation in pediatrics. Clin Pediatr 2018;57(4):481–3.

[65] Fareed S, Sarode N, Stewart FJ, Malik A, Laghaie E, Khizer S, et al. Applying fecal microbiota transplantation (FMT) to treat recurrent *Clostridium difficile* infections (rCDI) in children. PeerJ 2018;6:e4663.

[66] Hourigan SK, Oliva-Hemker M. Fecal microbiota transplantation in children: a brief review. Pediatr Res 2016;80(1):2–6.

[67] Paramsothy S, Paramsothy R, Rubin DT, Kamm MA, Kaakoush NO, Mitchell HM, et al. Faecal microbiota transplantation for inflammatory bowel disease: a systematic review and meta-analysis. J Crohns Colitis 2017;11(10):1180–99.

[68] Fang H, Fu L, Wang J. Protocol for fecal microbiota transplantation in inflammatory bowel disease: a systematic review and meta-analysis. Biomed Res Int 2018;2018:8941340.

[69] Nusbaum DJ, Sun F, Ren J, Zhu Z, Ramsy N, Pervolarakis N, et al. Gut microbial and metabolomic profiles after fecal microbiota transplantation in pediatric ulcerative colitis patients. FEMS Microbiol Ecol 2018;94(9).

[70] Qazi T, Amaratunga T, Barnes EL, Fischer M, Kassam Z, Allegretti JR. The risk of inflammatory bowel disease flares after fecal microbiota transplantation: systematic review and meta-analysis. Gut Microbes 2017;8(6):574–88.

[71] Sha S, Liang J, Chen M, Xu B, Liang C, Wei N, et al. Systematic review: faecal microbiota transplantation therapy for digestive and nondigestive disorders in adults and children. Aliment Pharmacol Ther 2014;39(10):1003–32.

[72] Xu MQ, Cao HL, Wang WQ, Wang S, Cao XC, Yan F, et al. Fecal microbiota transplantation broadening its application beyond intestinal disorders. World J Gastroenterol 2015;21(1):102–11.

[73] Kang DW, Adams JB, Gregory AC, Borody T, Chittick L, Fasano A, et al. Microbiota transfer therapy alters gut ecosystem and improves gastrointestinal and autism symptoms: an open-label study. Microbiome 2017;5(1):10.

[74] Zhao H, Shi Y, Luo X, Peng L, Yang Y, Zou L. The effect of fecal microbiota transplantation on a child with Tourette syndrome. Case Rep Med 2017;2017:6165239.

[75] Baxter M, Colville A. Adverse events in faecal microbiota transplant: a review of the literature. J Hosp Infect 2016;92(2):117–27.

[76] Wang S, Xu M, Wang W, Cao X, Piao M, Khan S, et al. Systematic review: adverse events of fecal microbiota transplantation. PLoS One 2016;11(8):e0161174.

[77] Woodworth MH, Carpentieri C, Sitchenko KL, Kraft CS. Challenges in fecal donor selection and screening for fecal microbiota transplantation: a review. Gut Microbes 2017;8(3):225–37.

[78] Paramsothy S, Borody TJ, Lin E, Finlayson S, Walsh AJ, Samuel D, et al. Donor recruitment for fecal microbiota transplantation. Inflamm Bowel Dis 2015;21(7):1600–6.

[79] Petrof EO, Khoruts A. From stool transplants to next-generation microbiota therapeutics. Gastroenterology 2014;146(6):1573–82.

[80] Tvede M, Rask-Madsen J. Bacteriotherapy for chronic relapsing *Clostridium difficile* diarrhoea in six patients. Lancet 1989;1(8648):1156–60.

[81] Carlucci C, Petrof EO, Allen-Vercoe E. Fecal microbiota-based therapeutics for recurrent *Clostridium difficile* infection, ulcerative colitis and obesity. EBioMedicine 2016;13:37–45.

[82] Khanna S, Pardi DS, Kelly CR, Kraft CS, Dhere T, Henn MR, et al. A novel microbiome therapeutic increases gut microbial diversity and prevents recurrent *Clostridium difficile* infection. J Infect Dis 2016;214(2):173–81.

[83] Seres Therapeutics Inc. Seres therapeutics announces key findings from SER-109 phase 2 study analyses; 2017.

[84] Orenstein R, Dubberke E, Hardi R, Ray A, Mullane K, Pardi DS, et al. Safety and durability of RBX2660 (microbiota Suspension) for recurrent *Clostridium difficile* infection: results of the PUNCH CD study. Clin Infect Dis 2016;62(5):596–602.
[85] Kokai-Kun JF, Roberts T, Coughlin O, Le C, Whalen H, Stevenson R, et al. Use of ribaxamase (SYN-004), a β-lactamase, to prevent *Clostridium difficile* infection in β-lactam-treated patients: a double-blind, phase 2b, randomised placebo-controlled trial. Lancet Infect Dis 2019;19(5):487–96.
[86] Wang Q, Euler CW, Delaune A, Fischetti VA. Using a novel lysin to help control *Clostridium difficile* infections. Antimicrob Agents Chemother 2015;59(12):7447–57.

Index

Note: Page numbers followed by "*f*" and "*t*" refer to figures and tables, respectively.

I

K

L

M

N

Kolibri 360 GmbH
Ferdinand-Jühlke-Straße 7
99095 Erfurt, Deutschland
produktsicherheit@kolibri360.de